ROUTLEDGE LIBRARY EDTIONS:
GLOBAL TRANSPORT PLANNING

Volume 6

I0124811

FUTURE TRANSPORT POLICY

FUTURE TRANSPORT POLICY

K. J. BUTTON AND D. GILLINGWATER

Routledge
Taylor & Francis Group
LONDON AND NEW YORK

First published in 1986 by Croom Helm Ltd.

This edition first published in 2021
by Routledge
2 Park Square, Milton Park, Abingdon, Oxon OX14 4RN

and by Routledge
605 Third Avenue, New York, NY 10017

Routledge is an imprint of the Taylor & Francis Group, an informa business

British Library Cataloguing in Publication Data
A catalogue record for this book is available from the British Library

ISBN: 978-0-367-69870-6 (Set)
ISBN: 978-1-00-316032-8 (Set) (ebk)
ISBN: 978-0-367-72658-4 (Volume 6) (hbk)
ISBN: 978-0-36772-663-8 (Volume 6) (pbk)
ISBN: 978-1-00-315576-8 (Volume 6) (ebk)

Publisher's Note
The publisher has gone to great lengths to ensure the quality of this reprint but points out that some imperfections in the original copies may be apparent.

Disclaimer
The publisher has made every effort to trace copyright holders and would welcome correspondence from those they have been unable to trace.

Future Transport Policy

K.J. Button and D. Gillingwater

CROOM HELM
London • Sydney • Dover, New Hampshire

© 1986 K.J. Button and D. Gillingwater
Croom Helm Ltd, Provident House, Burrell Row,
Beckenham, Kent BR3 1AT

Croom Helm Australia Pty Ltd, Suite 4, 6th Floor,
64–76 Kippax Street, Surry Hills, NSW 2010, Australia

British Library Cataloguing in Publication Data

Button, K.J.
    Future transport policy.
    1. Transportation and state
    I. Title   II. Gillingwater, David
    380.5'068   HE193
    ISBN 0–7099–3225–1

Croom Helm, 51 Washington Street, Dover,
New Hampshire 03820, USA

Library of Congress Cataloging in Publication Data

Button, Kenneth John.
    Future transport policy.

    1. Transportation and state. 1. Gillingwater,
David. II. Title.
HE 193.B88   1986   380.5'068   86-6385
ISBN 0-7099-3225-1

Printed and bound in Great Britain by
Biddles Ltd, Guildford and King's Lynn

CONTENTS

Contents

LIST OF TABLES

List of Tables

LIST OF FIGURES

1 INTRODUCTION

The Idea

Transport is important in everyone's life. It provides the opportunity of enjoying a wide range of recreational and leisure activities as well as opening up a large number of employment options. For productive enterprises it provides the means of bringing together the various inputs required by modern industry and to distribute their outputs to their customers. It is also important in other senses, however, not directly related to people's movements or the transportation of goods. Transport is visible, noisy and dangerous; it takes up huge quantities of resources which must be diverted from elsewhere and it requires the existence of a large bureaucracy and administration to ensure its efficient running. Transport, also of course, is a major industry in itself providing employment for millions of people throughout the world.

Given its importance, its scope and the diverse ways in which transport both directly and indirectly affects our lifestyle, it is hardly surprising that a considerable academic and popular literature has grown up on the subject. (The latter has partly been stimulated by the considerable fascination that some modes of transport seem to hold for certain people — a fact which is difficult to refute given the number of train-spotters to be found squatting at the end of station platforms.) In particular, numerous books on transport economics and on transport planning have appeared in recent years — e.g., and this is just a small selection from the UK, Bell, Blackledge and Bowen (1983), Glaister (1981), Button (1982), Stubbs, Tyson and Dalvi (1980). These vary in their approach and each offers a different view of the contemporary problems encountered by decision-makers involved with transport provision. Equally, they differ in their attitudes towards what motivates the users of transport services. Most of these volumes provide a thorough and up-to-date discussion of the relevant economic or planning techniques, albeit with varying degrees of rigour, and generally they contain a substantial amount

1

of factual information and institutional detail. As complements to these texts, more detailed works have been published offering histories or assessments of policies applied with respect to individual transport modes (e.g. Nash (1982) on public transport and Starkie (1982) on motorways), undertakings or geographical areas (e.g. Barker (1974) and Collins and Pharoah (1974) on transport in London). Finally, a wide range of specialist journals have appeared which offer both a *pot pourri* of recent research findings and commentary and comment on developments in the transport sector.

These books, journals and periodicals, however, seldom provide anything approaching an overall assessment of transport policy. This does not mean that policy is ignored but rather that its discussion tends to be either mode specific, or, in more general volumes and textbooks, tagged on at the end of discussions of technique and theory. This book, in contrast, is concerned exclusively with transport policy. It makes no claims to be comprehensive but it does offer material which will partially fill this particular gap in the literature.

A discussion of transport policy and, in particular, some effort to predict the future patterns which policy is likely to take seems particularly apposite in the mid-1980s. Until a decade ago there had been a considerable degree of agreement over the general nature — if not always in the detail — transport policy should take and the objectives which were to be aimed for. The continuous economic expansion of the post-Second World War period and the acceleration of the growth of the consumer society in the 1960s induced a feeling of confidence and certainty about the future. Essentially, if there were problems associated with transport then society would soon 'buy' its way out of them as ever increasing amounts of resources became available and technical progress opened new horizons. There was also confidence that evolving techniques of economic control would provide a mechanism for evaluating and achieving macro- and micro-economic objectives. (The emphasis in the literature on technique and methodology is, therefore, understandable.) The situation has changed quite dramatically since that time.

The energy crises of the 1970s are often pointed to as major watersheds but, in fact, they represented only one element of the forces which caused the confidence of the 1960s to evaporate. Environmentalists were already at the time of the 1973 crisis pointing to increased problems generated by the widespread use of the petrol engine and to the broader aesthetic intrusions which were taking place and against which rising incomes were offering no protection. Third World countries, in complete contrast, were experiencing explosions in the sizes of their populations, and severe shortages of adequate transport facilities were becoming apparent. Changes in the nature of industry and the way people live have further contributed to the reassessments of transport policy which have emerged. Manufacturing industry in the developed world has changed in its composition, the service sector has increased in its significance while, more generally, there have been surges in urbanisation in the Third World and geographical spreading of cities in Western Europe and North America. The policies which had been assumed appropriate to adopt in the transport sector for the previous half-century suddenly became the focal point for fundamental criticism in the 1970s and 1980s.

Of course, transport policy is itself influenced by broader movements in overall political attitudes and economic and social policies. Again the past decade has witnessed quite marked shifts in political attitudes in many Western countries which have in turn caused people to rethink the approach which should be applied in the transport sector.

This book is concerned with examining both the influences which have stimulated recent policy changes as well as in describing the nature of the policy changes themselves. We are also rather more ambitious than this and attempt to extrapolate from the current situation into the mid-1990s. This may seem a rather speculative activity but, in fact, the task does not require quite as many assumptions and guesses as one might imagine. The future is not isolated from the present, nor, indeed, from the past, but represents an extension of it. This must be the case, as much for physical reasons as for any reasons of argument or of the automatic continuity of ideas. Transport is only possible because of the availability of a massive and sophisticated network of track and terminals — infrastructure which in some cases, such as ports and roads, has its roots sunk deep in history. Marginal changes are continually being made to this system but the resources required to replace it entirely are simply not available and nor are they likely to be in the foreseeable future. It must, therefore, be viewed as almost a constant in the equation of policy-making.

Even if one could sensibly consider completely replacing the infrastructure of transport, future policy is still likely to be heavily influenced by the past. Previous and existing policies are frequently retained for specific modes or to meet specified conditions quite simply because they are effective and are likely to be seen as eminently sensible over quite long periods. Other policies must continue because of commitments already made which are not easily removed. It is, of course, interesting and helpful to see why such policies come about — it offers clues as to the mechanisms of policy-making — but for predictive purposes the policies themselves are essentially fixed. Equally, other policies have been abandoned or modified to varying degrees in the past and the experiences of these changes, be they because the policies were ineffective or found to be politically divisive, are likely to make policy-makers think very carefully about whether it is wise to reintroduce them in the future. The past, therefore, provides lessons, and these are more often taken on board by future generations of decision-makers than is sometimes thought. If they are not, then antagonists may draw upon them for ammunition to aid their cause. Equally, an examination of policies which were considered but ultimately rejected in the past can offer helpful insights into policy-makers' implicit value judgements and goals. It can also provide guidelines as to where the real power lay in the decision-making process.

The Approach

One can approach public policy from a number of different directions and via several methods of analysis. The approach here is one of historical development, drawing more upon the insights of recent history and upon a careful extrapolation of influences to develop a picture of how transport policy is likely to change over the next decade or so. The time horizon is short, it certainly becomes difficult, even allowing for our earlier comments, to think much beyond ten years' time with technology advancing on so many fronts. (Who would have perceived the current scale of microcomputer use only a decade ago?) Essentially, therefore, we limit ourselves to considering how policy will develop with respect to a world employing more or less the same type of transport hardware that is available today. Where we do attempt to go beyond this, the suggestions and views are extremely speculative.

Our main preoccupation with the UK and US situations is both the result of needing a focal point for illustration and comparison and a reflection of our own backgrounds and expertise. These countries do, however, also provide good examples of policy development amongst the industrial nations but where other industrialised countries have different problems and pursue different priorities these are brought out. Equally, Third World countries often differ quite radically in their circumstances and these are reviewed. The conclusions reached, therefore, tend to be of a universal nature with due account taken where appropriate of national institutions and philosophies.

In a sense, of course, we are making some implicit assumptions about the distant future in discussing the way policy may develop into the 1990s. The longevity of transport infrastructure inevitably means that policy-makers themselves must have some picture of the developments likely to take place as we move into the next century and, certainly, they must take cognisance of the fact that changes will occur which may influence the long term viability of their ongoing investments. This longer term view on the part of policy-makers can be deceptive, however, and while it certainly exists most decisions involve either an implicit or explicit discount procedure which puts the emphasis on the near rather than distant future. Nevertheless, there is some evidence that policy makers are becoming increasingly concerned about wasting current resources by investing in technologies that may become obsolete very quickly. There is certainly more concern today to ensure that a high degree of flexibility is introduced into the policy-making process (i.e. programme planning) and into the design of the infrastructure provided (i.e. scope for extending its capacity at a relatively low cost). Some allowance has been taken of this in our thinking.

Our approach to transport is also thematic and deliberately avoids specific modal studies as much as possible. Transport is one of those rare instances where inputs are often used to delineate the sector — e.g. the bus industry. This is rather odd. Most industries are described in terms of their outputs (i.e. we normally think of the pet food industry, not the offal, scraps and can industry) and there are good reasons for this (and not, in our example, just deference to our sensitivity to the feelings of domestic animals). Most inputs are used in a variety of industries usually in conjunction with different combinations of other inputs and frequently in different ways. Further, policies are normally aimed at outputs and not inputs. (Although there may be exceptions to this in the sense that the outputs of some industries form inputs into other industries.) For these reasons we have tended to couch our discussions in terms of what might be called the transport service industries providing, for example, urban transport services and international transport services. The specific modes (buses, trains, planes, etc.) then simply form inputs into each sector in combination with fixed infrastructure, labour, fuel, etc.

There is a further point to be considered. If one actually looks at the administration and control of transport in most countries (and, indeed, within such international organisations as the European Economic Community) then this is seldom divided up on a modal basis. Local or metropolitan authorites, for instance, have powers over a range of urban transport services and, in the UK at least, have a remit to co-ordinate policy across them. From this perspective, therefore, there is a further justification for avoiding an excessive concentration on modal divisions.

One final point regarding our approach. We have attempted to write a book offering the maximum degree of accessibility to readers. It is hoped that it will be of interest to a wide audience of planners, economists, political scientists and policy-makers but that it should also not be so cluttered with jargon and technique (especially mathematics) that it is outside the reach of a more general readership. We offer, therefore, a more literary volume with only the occasional recourse to a diagram or table where rather more information is important to our argument. Should readers wish to go into any aspect of the subject in more detail we have included references within the text although these have been kept to a relatively small number of key books and articles.

The Layout of the Book

As emphasised in the previous section the book attempts as much as possible to avoid dividing its discussions of policy along modal lines. Rather it looks at broad themes and issues. To this end there are eight main chapters.

Chapter 2 provides an historical background setting out in very general terms how transport policy has evolved since the cessation of hostilities at the end of the Second World War. The chapter looks at the broad threads of policy and attempts to isolate common themes but primarily it is descriptive, offering some historical content into which the more recent developments in transport policy may be slotted and against which projections of future trends may be set.

This historical discussion is followed by a far more detailed review of the direction of urban transport policy (Chapter 3). Urbanisation is advancing at a rapid pace in the less developed world while there are quite noticeable changes in urban form in many developed, industrialised economies. The chapter is lengthy, in part because of the complexity of some of the policy issues involved but also because urban transport debates are felt likely to be at the forefront of transport policy-making in the coming years. Also, unlike some other areas, there are quite marked contrasts here between the nature of the most pressing problems confronting the administrations of cities in the Third World and those confronting urban authorities in the West. Essentially, therefore, our discussions have to be divided to allow for this dichotomy.

Chapter 4 is concerned with specific issues which are likely to receive detailed attention from policy-makers over the next decade. These relate to questions of ownership of transport facilities and the regulation of transport use. Clearly, these topics are not entirely unconnected — ownership, for example, in itself provides a means of regulation — but for the sake of exposition they may be conveniently separated. In particular, there are important ongoing debates about the extent to which public ownership of transport facilities should be contracted or expanded. Equally, where there is public ownership there are key questions now being asked about the appropriate form of organisational structure to adopt, the management goals to pursue and the degree of political intervention which is justified. The way these problems are resolved is likely to have a major impact on the way the transport system is likely to develop until the end of the century.

Similarly, we are currently going through a major rethink regarding the nature and form of regulation most appropriate to achieve the public objectives set for the private transport sector. Traditionally, even in the most free market economies such as the USA, governments at both the central and local level (and federal where applicable) have exercised considerable control over the operations of privately owned passenger and freight transport modes. The last decade has witnessed in many countries a major revision in this situation. The regulatory regimes of many nations have been substantially reformed. Whether these changes are likely to be permanent or whether reaction will set in is likely to be a central theme in the way future transport provision develops.

Chapter 5 takes a specific look at the way policies are developing with respect to the international movement of goods and people. International transport is the most rapidly expanding market for transport services and this expansion is likely to continue into the foreseeable future. At present the system of international agreements designed to cope with this expansion is under considerable strain and, indeed, much traffic is carried on the basis of *ad hoc*, bilateral agreements. The pace at which greater international co-ordination and co-operation is achieved and, indeed, the form that it takes will affect both the pace at which demands for international transport will be met and the nature of the ultimate international transport system which emerges.

The links between transport and development are examined in Chapter 6. Transport policy has, rightly or wrongly, traditionally been perceived as a useful and potent tool for achieving national economic growth and for tackling specific regional problems of under-development. This approach, or at least the *carte blanche* nature of many policies in the past, has recently come under detailed review. While transport is likely to continue to be employed as an instrument in development and spatial economic policy in the future, the nature of the approach adopted is likely to change. It is the latter point which provides the stimuli for this antepenultimate chapter.

Transport plays an important role in providing access to a wide range of facilities for those who might otherwise find themselves disadvantaged by isolation. Equally, mechanised transport has wide implications for the non-user — it is a major intrusion in the natural landscape, it pollutes the atmosphere and puts the safety of pedestrians and others in question. Chapter 7 concerns itself with ways in which these wider social and environmental considerations are likely to exert an influence on transport policy in the 1990s. Certainly the 1960s and the 1970s witnessed a considerable upsurge in environmental, ecological and distributional matters but whether this concern is likely to continue in the future and the impact such lobbies are likely to exert on actual transport policy is a topic to be examined.

The final chapter differs from the earlier ones. It is really our only attempt at 'star-gazing' and then it is only on a very limited scale (both in terms of the scope of the chapter and the objectives which are set). We are concerned here with making some tentative projections of the longer term impact of new or radically modified technologies on the transport sector and on policy-makers' reactions to them. Our approach is conservative and any assumptions of substantial change are tempered by the fact that most of the supposed leaps forward in the past have turned out to be rather tortuous crawls. It is unquestionably true that technology will change but, and this is the key point, most successful technological innovations do not suddenly appear as manna from heaven but rather gradually evolve at a pace and in a form determined by prevailing economic, social and political attitudes. It is in this context that we approach our discussion.

When initially sitting down to write this book it was envisaged that there would be a concluding chapter bringing together all of the main policy developments discussed and setting them in some form of overall context. As we are sure reviewers will note, there is in fact no such chapter. Our views on this changed as we went along. The simple fact is that transport is not homogenous but rather represents a diverse set of industries, each with its own unique set of characteristics. There is no reason whatsoever to suppose that the policy adopted, say, with regard to the provision of urban bus services in Glasgow, is in any way going to be related to the policies pursued with respect to the optimal provision of international road haulage. Where there are overlaps (e.g. domestic attitudes towards regulation in general spill over into national attitudes towards the regulation of international transport) this is pointed out as we go along. In addition, each chapter offers summaries and conclusions relevant to its particular topic or theme. To attempt to devise an all-embracing set of conclusions regarding the direction in which policy is moving not only seems rather artificial but is, in fact, in direct conflict with the arguments being advanced.

References

Barker, T. C. (1974), *History of London Transport: Passenger Travel and the Development of the Metropolis: Vol. II the Twentieth Century to 1970* (George Allen & Unwin, London).

Bell, G., D. A. Blackledge and P. Bowen (1983), *The Economics and Planning of Transport* (Heinemann, London).

Button, K. J. (1982), *Transport Economics* (Heinemann, London).
Collins, M. F. and T. M. Pharoah (1974), *Transport Organisation in a Great City: The Case of London* (George Allen & Unwin, London).
Glaister, S. (1981), *Fundamentals of Transport Economics* (Basil Blackwell, Oxford).
Nash, C. A. (1982), *Economics of Public Transport* (Longman, London).
Starkie, D. (1982), *The Motorway Age: Road and Traffic Policies in Post-War Britain* (Pergamon, London).
Stubbs, P. C., W. J. Tyson and M. Q. Dalvi (1980), *Transport Economics* (George Allen & Unwin, London).

2 THE EVOLUTION OF TRANSPORT POLICY

Introduction

The evolution of transport policy has been marked by periods of continuity and change. Whilst there is some measure of agreement about the periods when transport policy remained reasonably constant, there is major disagreement about when changes took place and — much more importantly — why. This is hardly surprising. For example, we know that the automobile is a relatively new phenomenon. Even seventy years ago, it was something of a novelty, and identified usually with a particular, small and affluent group. Today, the automobile is ubiquitous and its growth sweeps inexorably upwards, transcending barriers of income and social class, even though to run one is, next to housing costs, the single most significant part of anyone's disposable income. But when did this change take place? Did it occur very slowly or did car ownership and use undergo some structural shifts, with ratchet-like effects? The basic answer is that we do not know, or rather the whole picture is so complex that a simple cause–effect type of analysis, where we try and isolate one or two key factors, leads us nowhere.

Does this mean then that an analysis of the evolution of transport policy suffers a similar fate? In part, the answer is Yes. There is certainly little agreement among economic historians. For many, transport policy is usually seen as a product of changes taking place in the economy. Many of their arguments are 'developmental' — that is rapid innovation in transport is equated with rapid economic development, so transport policy develops out of changes which have already taken place in the wider economy. In this sense, transport policy, as is public policy in general, is seen as inherently reactive.

However, since the 1960s and as a result of the search for evidence to support a strong link between transport innovation and economic growth, the so-called 'new' economic history has stood the development approach on its head (see Chapter 6). As Hart (1983) puts it, in certain respects improved transport is a

11

luxury afforded out of economic growth. Single innovations, like the automobile, are no longer seen by many economic historians as leading an economy inevitably towards 'take-off' or a 'drive to maturity'. Instead, emphasis is placed on the complexity of the origins and continuation of growth, with transport innovations being interpreted as only one aspect of economic gains arising in societies already favourably placed for expansion. In terms of transport, the implications are that policies were not only concerned with consequences but were also trying to facilitate changes in what were deemed to be desirable directions.

This contrast suggests that, if there is little agreement about the relationship between developments in transport and economic development, then it becomes even more difficult to expect agreement about the continuity and change in transport policy. A further complicating factor is the assumption that policy for transport is grounded in an assessment of the facts pertaining at the time. In other words, transport policy is born out of a reasoned analysis of the facts, allied to a clear and unambiguous objective about what the policy is supposed to achieve. For example, in a recent study of public control of the British bus industry, Glaister and Mulley (1983) appear bemused by the fact that the development of quantity licensing control was motivated more by political influences and administrative factors than by reference to any economic rationale.

> The Traffic Commissioners assumed at an early stage that it was part of their function to achieve uniform rates per mile not only in their own Areas but right across the country and they took some considerable trouble in order to achieve it. They carried out this duty *despite* the fact that the Act did not *explicitly* require them to strive for fares uniformity. The Act only required that they should not be 'unreasonable' and that 'where desirable in the public interest the fares shall be so fixed as to prevent wasteful competition'. We see this as another manifestation of the desire on the part of the authorities, *lacking an economic rationale*, to 'tidy-up' the industry. (Emphases added)

What this shows is quite interesting because behind the analysis is a particular 'view' of what the Traffic Commissioners should have done, based on an implicit assumption that the transport policy in the 1930 Act was somehow rational and that the Commissioners either chose to ignore it or assumed their own. This is one 'reading' of the events: that the intention of the Act was somehow distorted by those with responsibility for its implementation. However, as we know from the records and reports of the Traffic Commissioners, they operated this part of the Act based on what they deemed at the time to be a logical administrative rationale as to what was 'unreasonable' and what was in 'the public interest'. There was nothing explicitly stated in that Act to say that they had to operate according to some pre-determined economic rationale.

The purpose of beginning with the evolution of policy in this way is to highlight the controversy which is an inherent part of the baggage accompanying transport policy. It is important to realise that there is no necessarily perfect correspondence between policy and the facts upon which the policy is supposedly based. It is also clear that at base there is a fundamental tension in transport policy which permeates its contemporary history: the tension between policy arguments based on notions of allocative efficiency (essentially arguments about pricing policy and distributional effects) and political obligation (essentially arguments about the need for state intervention and regulation to protect the public interest). The battleground is a familiar one, reflecting the polar ends of a continuum deeply rooted in liberal representative democracy: the role of the free market versus the public interest vested in the state.

It is this tension which provides our starting point. Firstly, because it makes it possible to locate transport policy at any point in time somewhere along this continuum — and so enable us for example to discuss shifts away from one approach towards another. Secondly, because it gets us out of the trap of being cast as economic historians searching for precise periods of continuity and change. All that we offer is another but different variation on a theme which has already been played by a number of researchers (cf. Aldcroft, 1975; Plowden, 1973; Dunn, 1981). All we do is recast some of the important evidence about the evolution of transport policy, not for its own sake but rather as a recognition of its importance in understanding competing arguments about future transport policy.

In order to clarify this evidence, we have simplified historical analyses with a view to identifying the key periods in the evolution of transport policy. It is quite tempting to provide an historical overview simply by cataloguing in traditional style the key dates and events. But our approach is different because we contend that future transport policy will continue to be heavily influenced by its history — not only because of heavy sunk costs in, for example, infrastructure but more importantly because of the basic tension which exists. It is this tension which creates the fundamental dilemma for those interested in transport policy. Although many neo-liberals might agree with Ponsonby's (1969) assertion that public transport undertakings are a blunt, inefficient and ineffective instrument for redistributing wealth, such an instrument does not exist for its own sake, but rather because it is preferable to having no instrument at all. This is precisely the logic of the contradiction which many forget. For supporters of allocative efficiency, public interest arguments are a tiresome if necessary barrier to real economic advance and social progress, if only because they presuppose bureaucratic administration and some degree of interventionist planning. Similarly, for those ardent supporters of public interest, allocative efficiency arguments appear burdensome and light years away from improving the administrative allocation procedures which have been built up over many years by those in public service.

One final comment needs to be made, which concerns the focus of the chapter. It is heavily biased towards the UK and North American experience; only in passing will we make reference to other, mainly Western countries' transport policy. This has been done for three reasons: first, because precious little historical work has been undertaken into the evolution of transport policy in other countries, and even less is available in English translation; secondly, in a way transport policy in the UK and USA has been significantly in advance of policy elsewhere; and finally, there are fundamental differences in cultures, social structures and political philosophies. This is particularly significant in the context of a Common Transport Policy for the European Communities, and is worth dwelling on at this stage.

One of the major stumbling blocks currently facing the European Communities (see Chapter 6, pp. 145–51 for more detail) is the quest for a Common Transport Policy. A superficial analysis puts this down to a lack of political will to come to grips with the problem. A careful analysis, however, suggests other more complex processes at work, which hinge around differences in individual member states' cultural traditions, their perceptions of the role of public transport, and the place for intervention and regulation. Basically there is no common agreement, at least there is little in the way of shared views or a community of interest between, on the one hand, Britain and Eire and, particularly on the other, France and West Germany. Here we have fundamental differences which focus on the competing ideologies underpinning the respective political institutions. This boils down to the basic tension, noted earlier, between appeals for greater efficiency via the market mechanism (the 'Anglo Saxon' approach), and plans for furthering the protection of the public interest by ensuring the regulation of transport for social ends (the 'Continental' approach). For example, both France and West Germany tend towards the Continental approach, whereby transport policy is treated as an instrument in wider social policy formulation. Arguments about allocative efficiency are present, but in practice not deferred to. The French policy of *droit de transport* is perhaps the best example of this. In West Germany public transport is integrally locked into land-use planning. A good example is provided by the Rhine–Ruhr region where since 1920 the Siedlungsverland Ruhrkohlenbezirk (SVR) has been conducting a unique experiment. Because of the economic and social problems of the area, public transport policy is seen as important in creating an integrated system comprising the electrified S-Bahn inter-urban network of the German Federal Railway, new and more local urban railways, and new interchanges between the two rail systems, which link in with feeder bus systems. The key point is that the objectives of these transport initiatives are social, in the sense that the SVR is trying to create local concentrations of employment and high-density residential areas around transport modes, with a view to bringing as many people as possible within easy access to public transport (Hall, 1977).

We find therefore that both France and West Germany's political institutions take a radically different view of transport provision to those of the UK and USA. To say that they are wrong, or that their solutions are inefficient, is either a bad case of cultural myopia, or of being blind to a political and cultural tradition based on values and operating principles different to our own. As Weiner (1981) has argued, it is a tradition that we ignore at our peril.

The Four Ages of Transport Policy

It is possible to conceive of the evolution of transport policy as a process of four periods of continuity marked by three periods of structural change. The periods of continuity can conveniently be called 'The Railway Age', 'The Age of Protection', 'The Age of Administrative Planning', and 'The Age of Contestability'. The three periods of structural change cover (i) the decline of the railway and the rise of the bus and car, (ii) the consolidation of regulation and the expansion of state control; and (iii) the breakdown of state control and the rise of disengagement.

The Railway Age dominated the period from the 1800s until the early 1920s; the Age of Protection was dominant from the end of the First World War (1918) until the cessation of hostilities in 1945; the Age of Administrative Planning really took off in the mid-1930s, lasting until the mid-1970s; and the Age of Contestability characterised the period from the late 1960s up to the present time (and, arguably, into the future period with which this book is concerned).

If these are our periods of continuity, then the periods where structural changes were being wrought span the following years: 1916 to the early 1920s, with the rise of the Age of Protection; the mid-1930s to 1945, with the rise of the Age of Administrative Planning; and the late 1960s to the mid-1970s, with the advent of the Age of Contestability. Broadly, the argument runs that for a period of more than fifty years there was some general measure of agreement about the need for a formal transport policy, its objectives and content. In the 1950s, this consensus began to be questioned and eventually became the subject of open assault in the 1960s, with a consequent erosion of support and eventual breakdown by the late 1970s. What is remarkable about this periodisation is that it appears to apply to the evolution of transport policy not only in the UK and the USA, but also in Canada and to a lesser extent Australia.

The Railway Age

From a historical perspective, this particular era is one which has received the most attention from historians of transport and economic policy — and where controversy is never ending. (The issues are treated systematically in Gourvish (1980) for the UK, in Fogel (1964) for North America, and in a comparative study of American and European policy by Dunn (1981).)

From a transport policy point of view, the 125 years from 1800 is a remarkable one, if only because of its duration. This era was dominated by policy discussions and legislation directed at railways and rail operations, almost to the exclusion of all other in-land modes. Interestingly, the main focus of policy was not so much about what was being provided but rather who was providing it and how it was being operated. As in most countries where rail-based investment was taking place at a rapid rate, the main source of funding came from profit-seeking private investors. It was this aspect which provided the main focus for policy — the awareness of increasing monopoly power of the railway companies. In Britain this led to the Railway and Canal Traffic Act of 1875. Until this time, policy had been based on a concern for adequate safety standards; the 1875 Act introduced extensive controls over railway charges. Whether the railway companies were in fact abusing their monopolistic power is a matter for conjecture, but what cannot be denied is the Act's overall significance in terms of the evolution of UK transport policy.

Although the Railway Age has been described as an atypical period in which investment was channelled into a mode of transport demanding an unusually heavy commitment to fixed plant (Hibbs, 1982), it is nonetheless a benchmark whereby a number of significant seeds for future transport policy were sown. For example, and as the 1875 Act illustrates, it became legitimate to regulate transport provision in spheres other than where the State was already intervening in the public interest. In terms of competition policy, these fears led to a situation where mergers and amalgamations were positively resisted until the 1911 Railway Amalgamation Report.

This evolution of transport policy in the Railway Age reached its zenith in the period immediately following the outbreak of hostilities in 1914. As the economic historian Aldcroft (1975) remarks, it came as no surprise that the government immediately assumed control of the railways. Although they remained under private ownership, their day to day management was in the hands of a government-appointed Railway Executive Committee, an arrangement which had been organised more than forty years previously in the Regulation of the Armed Forces Act 1871. This point cannot be overemphasised — the importance which *mass* transport assumed in the strategic thinking of the military in the Railway Age.

The planning which was required to manage this situation was created by 1912. The services of the railway network were to be at the disposal of the government, subject to the private railway companies being compensated on the basis of their net income at an agreed date. Thus by 1914 we saw the advent of a regulatory regime which had not been seen in any guise before: a complex array of regulatory devices covering quantity and quality controls, pricing and charging systems, and state control through administrative planning.

By 1918, and the cessation of hostilities, the Parliamentary Select Committee on Transport was able to provide a glowing account of the administration of the railways during the crisis years: despite a significant loss of manpower to the armed forces and a smaller rolling stock, the state organised railways had carried a much larger volume of traffic than before the war.

However, this was not achieved without cost. Although compensation was paid to railway companies on the basis of their net income for 1913, a government committee reported in 1921 that the outstanding liability amounted to £150 million, mainly for maintenance and repair. But perhaps the real cost was the lack of investment and re-investment during the war years; an estimate suggests that the level of investment was only half that of the pre-war period.

In the immediate post-war era, transport policy was concerned with dismantling the administrative planning system rather than with the physical state of the network or the problem of reinvestment. That said, the government actively considered a policy option which would have taken the railways wholly into public ownership. The war experience had provided an impetus for this, but perhaps more significant was the evolution of state intervention in the industry in the preceding fifty years. The argument put forward in 1918 by the Parliamentary Select Committee on Transport stressed on the one hand the need for unification of ownership in order to achieve a unified operation and, on the other, the need to achieve significant economies, given the industry's financial state at that time. However, the nettle was not grasped — unlike in France where the railways were taken wholly into public ownership on the basis of very similar arguments.

If the First World War years were the zenith of the Railway Age, then its nadir was approaching rapidly. The crisis tendencies noted earlier became so accentuated that eventually a set of policies were accepted (one of which saw the birth of a separate Ministry of Transport in 1919) to tackle them. However, this took time and it was not until 1921 that the measures were enacted in the Railways Act. In the mean time freight rates and fares continued to be controlled at their pre-war levels, though costs had risen by more than 100 per cent. Even though charges were revised and increased by a newly formed Rates Advisory Commitee, the result was a heavy loss in 1921, which the government was obliged under its wartime agreement to meet.

From a policy viewpoint, the 1921 Act was important for two main reasons. First, it introduced a scheme to merge the then 120 separate companies into four large groups. The aim was to eliminate rivalry and reduce duplication of facilities. There was also an assumption that significant economies would be forthcoming, including the standardisation of equipment and rolling stock. Secondly, it revised the regulatory system which controlled charges. The 1921 Act proposed a new Railway Rates Tribunal, with rates to be based on the principle of yielding a 'standard net revenue' equivalent to that of 1913.

What the Act did not change was the plethora of legal restrictions which effectively governed how the railway companies were to operate. Almost all of these were nineteenth century in origin and applied at a time when it was claimed railways had monopoly power. However, what was appropriate in the 1870s was demonstrably not so appropriate fifty years later. The first reason we have already noted — the coming of the Great War, and the way in which government effectively took complete control of the major inland transport system. A second reason was the Great War's impact on the financial fortunes of the multitude of railway companies.

However, there is a third reason which had perhaps greater impact. The wartime administration gave a whole new impetus to two innovatory forms of transport which were to provide the first real competitive test for the railway companies. These were road transport and aircraft. In 1918 it has been estimated that less than 350,000 motor vehicles were in use in the UK; within three years that figure had increased to almost 850,000 vehicles. So at the very time when the railway companies were struggling to reorganise themselves financially, a wholly new and quite different set of transport systems was being spawned. It is perhaps no coincidence that government transport policy began to move with these changes. Indeed, we have already seen that by 1919 a new government department was created to cope with these consequences — the Ministry of Transport — which would actively promote road transport, whilst at the same time incorporate the Railway Department set up in 1840 under the Board of Trade. The objective of the new policy was to promote road transport and encourage the most efficient combination of rail and road use.

The Age of Protection

We have explored the key points in the Railway Age, from the viewpoint of transport policy, placing particular emphasis on the regulatory side and the impact of the Great War. The Age of Protection has its roots deeply entrenched in the experience gained and consequences flowing from the latter part of the Railway Age, and the structural changes that were wrought then.

From a historical point of view, it is only relatively recently that much work has been undertaken on this period. Many writers prefer to use the term 'the inter-war period'. The point to note, however, is that there is nothing particularly transport-related, nor indeed any real historical justification for labelling this period in such a neutral way. From a transport policy viewpoint, it is clear that the era was dominated by arguments and debates about the need for protection. Indeed the title 'The Age of Protection' is attributable directly to Hibbs (1980), who asserts that it was the age of safety-first when a regulatory system was devised 'to be acceptable to both the bureaucrats and capitalists', in other words, a policy more in keeping with public-interest type arguments rather than the quest for allocative efficiency.

That the period from about 1918 until 1945 can be seen as a period of continuity might at first sight seem strange, given the massive structural changes which took place — for example, the Great Depression, the rearmament programme and the Second World War, coupled with little short of an explosion in the growth both of suburban-style living and motor vehicles. The continuity was basically one of transport policy, for throughout this period we can identify a process whereby successive governments tried to grapple with the consequences associated with these trends.

To give some idea of the magnitude of these changes, consumers' expenditure on all forms of transport between 1920 and 1938 in the UK increased by almost 75 per cent in real terms. Within this figure, expenditure on private transport increased by over 424 per cent and public transport expenditure increased by over 25 per cent. However, within the public transport sector, the major growth was in road passenger transport (over 523 per cent) with the railways' increase amounting to 7 per cent over the 18 year period. By 1938, expenditure on road passenger transport was almost double that on rail transport. At the start of the period, private transport accounted for an equivalent of about 18 per cent of public transport expenditure; by 1938 it had increased to over 75 per cent. The picture within public passenger transport also changed quite dramatically over the period. In terms of passenger miles rather than expenditure, the increase was of the order of 46.7 per cent, with most growth occurring in the period 1920 to 1929. In 1920, rail travel accounted for 59 per cent of all passenger miles, followed by trams and trolley buses, buses and coaches and other modes. By 1929 the picture was changing quite dramatically: the railway's share fell to 46.5 per cent, trams and trolley buses fell to just over 23 per cent, bus and coach travel increased to almost 28 per cent, and other modes fell to 2.5 per cent. By 1938, rail travel had fallen to 42 per cent of all passenger miles, trams and trolley buses to 17 per cent, while buses and coaches had increased dramatically to 40 per cent, and other modes to a little over 1 per cent. Although no figures are available for private transport, Aldcroft (1975) suggests that a figure of about 27,000 million passenger miles is probably on the high side, but enables a broad comparison to be made with the total of 47,794 million passenger miles for public transport.

Although no directly comparable figures are available for other countries, it is clear that similar forces were at work, leading to similar consequences for transport policy. In the case of North America, the picture painted above applies only generally to the more urban areas and particularly to the North Eastern Seaboard (Gottmann, 1961).

Broadly speaking the policy response to these changes was to try to bring some order to the situation. What is particularly interesting is that this response was adopted by virtually all governments whose countries were experiencing similar events — the United Kingdom and the United States were the first to seek to impose explicit controls over transport, followed notably by France and Germany. The common denominator in all policies was the assertion that transport was too important to be left to the market place. This assertion, almost without exception, was clearly justified on the grounds of protecting the public interest. Returning to the themes introduced earlier, it is clear that government regulation was seen as a legitimate and necessary device to control the exigencies which were beginning to emerge over the period: competition between modes, particularly within public transport, was perceived to be a 'public bad' which had to be eliminated; state controls over operations and ownership were therefore a logical outcome of such a policy objective; given the dramatic rise in vehicles and their use, and the demand for transport which was being expressed, it was vital that the issues of congestion had to be tackled but particularly public safety had to be protected.

Regulation in the name of protection became the clarion call. As Glaister and Mulley (1983) put it, the economic and political atmosphere was heavily anti-competition, favouring mergers, rationalisations and protection from cut-throat competition. This was based in part on fears of overcapacity and instability in the public transport industry, engendering a feeling that price cutting between hundreds of small independent operators would not be in the public interest.

In Britain this philosophy took root in a whole range of legislation. In this period a vast array of regulatory controls was introduced or extended to cover rail transport (1921), quantity control of private bus operations (London only) (1924), national quantity and price control over all bus operations (1930), public ownership of London's buses and underground railway (1933), quantity control of road haulage (1933), and — as a prelude to the Age of Administrative Planning — public ownership of road passenger and freight operations in Northern Ireland.

In the United States a similar process had been at work. Like the UK, the railways in the USA had been heavily regulated from the mid-nineteenth century onwards. Now regulations were introduced to cover all bus operations (at state level) (1925), coastal shipping

(1933), interurban bus operations (at federal level) (1935), road haulage (1935), airlines (1938), inland waterways (1940) and freight forwarders (1942).

Until 1935 most regulatory controls in the USA had been devised at state level, largely in response to the scale of mergers and rationalisations which had been taking place throughout the 1920s. Although early regulatory controls emphasised safety, increasingly control of entry and fares became the key focus of attention. The procedures adopted were quite novel when compared with the UK, because they were based on an assumption that public transport was to all intents and purposes a 'public utility', like electricity supply. The guiding hand was seen to be best provided by a paternalistic quasi-natural monopoly regulated by a State Commission. As a recent report of the Interstate Commerce Commission's Office of Transportation Analysis put it, 'stability in operating conditions through limiting entry of potential competitors was thought to be necessary'.

The Motor Carrier Act of 1935 brought federal regulation for the first time. Existing operators were given what have been called 'grandfather' operating rights, as established operators. Any new applicant had to demonstrate that what they proposed was in the public interest, spelt out as 'the present or future public convenience and necessity'.

However, by the time this legislation was enacted, the Greyhound Corporation had already established a national route network, following its history of merging and consolidating local bus companies, and was given grandfather authority. It was to all intents and purposes in a monopoly position prior to the 1935 Act. The new Act, therefore, ensured its protection from any major new competition.

The Public Utility Holding Company Act of 1935 had a devastating impact on public transport in the USA. The key part of the Act stated that any registered public utility holding company would have to limit its operation to a single integrated public utility system. The main thrust of this was to seek to regulate those companies with monopoly powers over a range of sectors and interests. At that time, particularly in the power industry, there were concerted attempts to adopt a measure of vertical integration, whereby a holding company could control everything from water supply to electricity generation to a public transport company, using electricity as a means of propulsion. This Act attempted to regulate that type of control. In fact this was in part a consequence of a previous federal policy which, in 1919, set up the Federal Electric Railway Commission to publicise the benefits of electric traction. The reason for this was rooted in the consequences of the First World War, whereby US policy had, like Britain, been based on controlling fares and rates of public transport. Unlike Britain, however, the US government had not entered into an extensive programme of financial support for the industry. As a result, due to rapidly increasing operating costs and fixed rates, one third of public transport operators had gone bankrupt. This situation, combined with the established companies' reluctance to provide the necessary capital to reinvest in electric power, led to many public utility companies producing electricity to buy their way into public transport operations by buying out the horsecar owners. By 1931 about 50 per cent of public transport companies and over 80 per cent of total revenue passengers travelled on routes controlled by these holding companies.

When the legislation was being debated there was no reference to its potential impact on public transport provision. When it became law, each holding company basically had to chose between their various operations. For example, the Cities Service System, an Ohio holding company with interests in power supply, public transport and petroleum, appealed to the courts but to no avail. They chose to become a gas and petroleum operation. Within a few years most holding companies had divested themselves of their public transport operations.

In Britain the approach to regulation was based on a similar logic but with different policy instruments. There was also no equivalent to the impact of the Holding Companies Act. The 1930 Road Traffic Act marks the most important piece of formalised transport policy in this period, it is also firmly within the containment tradition of public policy — that is an attempt to place restrictions on the use and operation of motor vehicles to meet particular policy objectives.

Glaister and Mulley (1983) have reconstructed the evidence which led up to the 1930 Act, and take the view that it was concern for public safety in the early 1920s which provoked a policy response. Within a few years of the cessation of hostilities, the number of vehicles on the road had increased by an estimated 143 per cent; by the end of the 1920s the increase was of the order of

471 per cent. This generated a wholly new pattern of transport, and exacerbated problems of congestion and accidents. The reduction in the operating costs of buses, attributable to technological advances in engine design and the introduction of pneumatic tyres, all served to provoke considerable public concern.

These factors, coupled with the newly reorganised railway companies' attempts to seek powers to enter road passenger transport in direct competition with bus operators, led the government to set up a Royal Commission on Transport in 1927. The terms of reference embraced the public interest philosophy.

The Commissioners reported in 1929 and their Report became the basis of the 1930 Act. The two most contentious problems which the Report addressed concerned the regulation of public service vehicles: firstly, who would administer any new legislation and, secondly, the best means of regulation to achieve the aim of co-ordination.

The dilemma which the Commissioners faced was nothing new. It was how to exercise a uniform system of control and yet retain sufficient flexibility to reflect local conditions. Their solution was to advocate a system of local Traffic Areas administered by government-appointed local Traffic Commissioners with (i) licensing powers covering vehicles, levels of service and employees, and (ii) powers to regulate entry into the industry, whereby the established operator was given priority over any new applicant. The issue of competition, which was incorporated into part (ii) of the proposed Traffic Commissioners' powers, was examined in depth. The influence of American opinion and evidence was clearly very strong indeed, and rested on an unequivocal pronouncement by the Supreme Court of the State of Pennsylvania that 'unrestricted competition in such utilities has been, by experience, definitely shown to be ultimately unwholesome for the community. Unrestricted competition and regulation are inconsistent'.

The Royal Commission's proposals were enacted one year later with a Bill which appeared to capture the public mood in favour of restriction. However, one reason why this was the case has probably as much to do with the presentation of the evidence as with the facts themselves, since most of the evidence came from the railways and established operators rather than small bus companies and users of public transport. That said, the Bill became law in 1930, which set down the basic regulatory regime for the next fifty years. Three years later a similar Bill for the control of road haulage was enacted, with almost identical but more limited provisions — the Road and Rail Traffic Act of 1933.

So what was the effect of this approach to transport policy in the Age of Protection? Broadly speaking, the policy objectives appear to have been fulfilled. Aldcroft (1975) asserts that its impact was to freeze the pattern of development in road transport, the growth of the industry was checked, both in terms of numbers of operators and vehicles. There is however no evidence that the longer established operators abused their privileged position under the licensing system. According to Aldcroft, 'passengers not only enjoyed the benefits of safer and more reliable services but also the advantages which followed from a much greater degree of co-ordination of timetables and fare schedules'. The same could also be said about the impact of regulation on road haulage.

Under the 1930 and 1933 Acts, the regulatory bodies strove to provide a set of conditions designed to cope with the exigencies they faced. They were little more than passive licensing authorities, their main weapon being an administrative one of withholding or not granting an operator's licence if certain conditions were not fulfilled. They had no powers to compel any operator to provide a service. It is clear that the system relied heavily on the goodwill of both operators and the Traffic Commissioners. By 1938 it seems that the approach to licensing had become standardised, and the processing of licences a routine activity.

Within twelve months Britain was again at war, and the continuity in policy which had characterised the turbulent period through post-war reconstruction, international economic depression and developing prosperity was put to the utmost test. As was the case immediately prior to the First World War, considerable thought had gone into preparing for hostilities, and the experience gained in the 1914–18 period was rapidly brought to the forefront. As in 1914, many of the plans and programmes had their roots in earlier transport legislation and transport organisation. The Railway Executive Committee was brought into effect once more, on terms similar to those which had nearly crippled the industry only two decades previously. In addition, three administrative innovations from the inter-war period proved particularly significant. The first followed the implementation of the London Passenger Transport Act of 1933 which created a public operator, London Transport. The experience gained of combining and co-ordinating separate enterprises into one transport undertaking under public control was salutary. The second was the role of the Traffic Commissioners. They played an instrumental role under the emergency powers, since they had the ability not only to grant or withhold licences but to issue permits at their discretion. This created the flexibility in planning passenger transport which was so sorely missed in the 1914–18 war. In effect public transport was taken into public control through the process of requisitioning, as indeed was road haulage. However, it was a third piece of legislation which was of greatest significance, since it formed the prelude to the emergence of the new era in the evolution of transport policy — the Age of Administrative Planning.

The experience gained in implementing the London Act was paralleled by an equally significant Act for transport in Northern Ireland — the Road and Railway Transport (Northern Ireland) Act of 1935. This created the Northern Ireland Road Transport Board, with powers to acquire compulsorily virtually the whole of road passenger and freight operators in the province. This Act together with the Railways Act of 1912, which applied to Northern Ireland, and the fact that Belfast City Transport was a municipally owned operation, meant that public transport in the Province was in effective public ownership from the early 1930s. The importance of this for the new war effort was unmistakable. It had highlighted the administrative mechanisms needed for a co-ordinated approach to planning. In many respects the greatest contribution to the evolution of transport policy made during the Age of Protection was not the system of regulation set up under the 1930 Act, but the way in which it enabled the seeds to be sown for future transport policy.

The Age of Administrative Planning

The Age of Administrative Planning began in the early 1930s. Like the previous eras, it began under conditions of turmoil and came to something of a peremptory end in a period of turbulence and uncertainty. In a way it is possible to stitch together the Railway Age, the Age of Protection, and the Age of Administrative Planning into a sort of *longue durée* of transport policy, where arguments predicated in the public interest dominated those of allocative efficiency. The Age of Contestability which followed appears to have ruptured that philosophy.

It is often taken for granted that nationalisation was an inevitable outcome of the first post-war election. What we have shown is that its roots go much further back, and that the motivation for public control was based on a number of particular events. After all, Britain unlike the United States was the only country to pursue public ownership through nationalisation, and yet the United States pursued its own version of administrative planning with different outcomes.

The starting point for the Age of Administrative Planning has to be the advent of the war in Europe and the war with Japan. Both Britain and the United States, together with all those countries directly engaged in the war effort, chose the path of a formalised administrative planning approach to transport policy, corresponding very closely to the military chains of command. Without exception, rationing became the dominant process of allocating scarce resources between competing users. From a transport point of view, the major impacts were in terms of fuel rationing, a shortage of vehicles and spare parts for private use, the requisitioning of public transport carriers for military ends, and the staggering of working hours for those engaged in industrial production. The combined effect was to boost dramatically the use of public transport. In the United States, for example, the number of passengers climbed to about 19 billion (US), almost twice its pre-war level. As most commentators now accept, the 1939–45 war provided the conditions both for a structural change in transport, leading to a different type of transport policy, but it also represented a temporary hiatus in the inexorable process of the gradual dominance of private transport over public transport.

In terms of the development of transport policy, three major factors dominated its foundations: the explosion in private transport, especially the ownership and use of the private car; the continuing financial plight and operational weakness of the railways; and the phenomenon of urban growth and change, and the transport problems which this created.

These factors were clearly visible towards the end of the Age of Protection, but during and immediately preceding the ending of hostilities they provided the backdrop against which reconstruction would have to take place. It is no surprise therefore to find that transport policy emerged as a combination of pre-war policy, coupled with a new style which emphasised planning as an additional form of control. Those responsible for transport policy were consciously aware of what had happened following the 1918 armistice and concerned that the same problems would not recur. To oversimplify events, the policy which emerged at the start of this period effectively ignored the consequences of the growth of private car ownership, and focused instead on solving the railway problem and providing a comprehensive and integrated approach to the provision of a public transport service. It is interesting that this was the common response of each one of those countries which had been involved in the war effort.

What this policy approach meant in practice was to break the transport problem down into three component parts, involving three quite different but supposedly related policy responses. Firstly, to restructure railway operations nationally via increased state involvement; secondly, to focus particularly on the structure and organisation of public transport in predominantly heavily urbanised localities and big cities; finally, to see transport networks and infrastructure as a means of reorganising urban growth in a systematic way. Each of these became an additional layer on top of the mainly regulatory based transport policies which were already in existence. This approach to solving complex social, economic and political problems was based on the notion of administrative planning, with antecedents in the Age of Protection.

From the evolution of transport policy perspective, the beginning of the Age of Administrative Planning had its foundations firmly rooted in public interest type arguments. Administrative planning was seen as a logical extension of earlier experiments in state intervention in transport; as Aldcroft (1975) notes, the experience of two world wars had made the idea of unified ownership and control, vested in the state, a much more acceptable proposition than it might otherwise have been. Britain's experience was in principle no different from that of other countries, like the United States, Germany or France. Although there were differences in administrative style and the degree to which legislation was used, the outcomes were remarkably similar.

In Britain the response was to create a British Transport Commission (BTC) with administrative responsibility for the planning and control of transport operations on a day-to-day, year-by-year basis. The other two strands were to set up a statutory planning system to control land use through the preparation of long-term development plans, and a regulatory regime to control the spatial distribution and location of manufacturing industry. Britain then relied heavily on legislation to pursue a transport policy, which spilled over into other areas of government responsibility. The Cabinet, through its system of committees, became the key administrative device where coordination between these planning centres took place.

Both Germany and France pursued similar approaches, particularly France. Germany, under the influence of American public administrators, adopted a federated solution: the new federal government effected the administrative planning and control of strategic public transport (mainly the railways and inter-urban public transport) and distribution of industry policy, with the *Lande*, or states, having responsibility for land use controls and local public transport. The relationship between federal and state governments was primarily financial, of grant aided assistance, and regulatory, through the equivalent of a federal trade commission.

The United States, and latterly Canada, concentrated much more on the legally based regulatory side of administrative planning, with *ad hoc* planning commissions being created at district and state levels. These generally had few if any statutory powers, being purely advisory, and were concentrated in the big cities. Day to day operational control became, on the one hand, concentrated in locally regulated and often municipally owned organisations and, on the other, large regulated monopolies (like Greyhound and Trailways) subject to the Interstate Commerce Commission, and ultimately to the Federal Trade Commission.

In the meantime the real world of transport use and provision was being turned upside down. The growth of private car ownership was accelerating apace: in road haulage, own-account operations were becoming an increasingly attractive proposition for firms; and physical reconstruction of war damage was placing an enormous strain on public expenditure and private sector resources, creating conditions of scarcity. As a result the transport problem, which had been identified in the late 1930s and moulded into policy by the middle 1940s, intensified. Scarce public sector resources needed for improving and replacing vehicles and infrastructure now in the domain of the state were in direct competition with declared priority needs in other public policy areas, and the growth in car and goods vehicle ownership was creating capacity problems for the road network, with increasingly vociferous demands to invest in new roads on the basis of a national strategy. The patronage of public passenger transport reverted to a pattern of pre-war stability followed by gradual decline. Increasingly transport assumed an even larger proportion of total expenditure and investment, reflecting the fact that by 1970 on average people were making more than double the number of journeys they made in 1950.

In the United States the picture was very similar. Total passenger miles between 1959 and 1970 increased by 135 per cent. Of this, the share taken by the private car remained constant at 87 per cent, whereas public land transport patronage fell by 36 per cent. In the period, air transport increased its share from 2 per cent to 10 per cent. Schnore (1965) subsequently demonstrated that public transport was much more responsive to areas of high population density, particularly those in the older and larger cities. The newer medium-sized metropolitan areas and suburbs of older cities showed a much lower propensity to use public transport. The most rapid decline had occurred in the smaller, lower density and younger cities where the private car had already asserted its adaptability.

By the late 1960s and early 1970s, and for the first time in our history, a whole generation had been brought up on a mode of transport which did not in any way rely upon public transport. As Solomon and Saltzman (1971) put it, the public transport riding habit was gradually dying off with its passengers. In 1950 less than half the number of households in the United States owned a private car; twenty years later only 15 per cent did not, and 35 per cent owned more than two cars.

By this time and for these reasons a number of cracks had begun to appear in the edifice of the administrative planning regime. Five of the more important ones were:

(i) the continued preoccupation of transport policy with public transport operations;

(ii) a belated recognition that the basic principles of this approach to transport policy in the Age of Administrative Planning had not been worked out in sufficient depth;

(iii) the consequences of the increase in road traffic, the sudden emergence of extensive road building pro-

grammes in the late 1950s, and the environmental effects
of such schemes;

(iv) the belief that road building was an essential corollary of
policies to promote national economic growth; and

(v) the increasing extent and hidden nature of subsidies
and support to public passenger transport from public
funds.

In 1962 the British Transport Commission was abolished,
although its powers remained vested in other public agencies. In
1964, however, the Urban Mass Transportation Act was passed
in the United States, this being the first significant piece of
legislation aimed at improving public transport since the
emergence of the Age of Administrative Planning. This Act had
explicit social objectives, providing Federal grants for two-thirds
of the net cost of public transport projects and also for research,
development and demonstration projects. By 1969 capital grants
and loans of almost 600 million dollars had gone to 28 states,
mostly in the purchase of new equipment and the upgrading of
their operations.

In Britain the formative equivalent came in 1968 with the
passing of a new Transport Act. Basically this Act brought
about a major reorganisation of public transport management. It
was clear that the main intention was to try to reverse the trend
towards commercial freedom and the disintegration of public
transport services. The mainspring was still administrative
planning in the public interest, to ensure that the transport
system took full account of the social and economic requirements of the country as part of a coherent and integrated
whole. The proposals followed broadly the pattern of the 1947
Act: the first section redefined the size and shape of the basic
railway network in the light of changing circumstances; the
second envisaged a reorganisation of freight traffic, creating a
new umbrella holding company; and finally, an extensive
reorganisation of public transport was proposed, with Public
Transport Executives for each of the major urban areas and a
National Bus Company to control the rest.

Both the Railways Board and the Freight Corporation were
required to promote 'the provision of properly integrated services
for the carriage of goods . . . by road and rail', and the
reorganised bus operators had to promote the co-ordination of

local services. For the Passenger Transport Executives, their main duty was

> to secure or promote the provision of a properly integrated and efficient system of public passenger transport to meet the needs of that area with due regard to the town planning and traffic and parking policies of the councils of constituent areas and the economy and safety of operation [See Chapter 7 for details.]

In parallel to these events was the continuance of the enthusiasm for new road building. Roads and road-building had always been regarded as somehow not part of transport policy, but traffic policy. Although this goes back to the organisation of the Ministry of Transport in 1919, the important point to note is that roads had their own legislation and financial programmes under the Highways Acts. In fact it was not until 1973 that the two parts of policy were brought together for the first time, in the creation of Transport Policies and Programmes. A similar situation prevailed in the United States, where transport and traffic policy never became common partners in the Department of Transportation. It was not until 1956 that the Federal National Highway Trust Fund was set up. This provided up to 50 per cent reimbursement to those states constructing primary and secondary roads, together with 90 per cent for the building of a new national motorway network — the National System of Interstate and Defense Highways.

By the late 1950s in the United States the phenomenon of the land use transportation study had really taken off, in a concerted attempt to bring not political but a 'scientific' rationality to a complex social problem: computer-based mathematical modelling of the relationships between land use, traffic movements and forecasts of population growth became part of the conventional wisdom of the Age of Administrative Planning. By the early 1960s the phenomenon had ventured into European territory, with all major urban areas without exception being subjected to this kind of study.

The impetus for this came from two main sources: official forecasts about the increase in population; and a belief in the view that existing urban areas were heavily congested, both physically and economically. The result was therefore to seek solutions to what were seen as capacity problems, requiring supply-led initiatives by the public sector. The solutions preferred were transport and infrastructure led — a combination of new free-standing cities in the British New Town tradition, coupled with massive investment in new motorway construction to complete a national network and canalise traffic away from and around existing established urban areas. The massive land use transportation studies provided the justification for the spending programmes.

Two events of major significance punched a very large hole in this scientific approach to planning in the UK. The first was brought about following a recognition of the potential impact of road building on the existing urban fabric. At the public inquiry into the *Greater London Development Plan* in 1968, which maintained the need for new ring-and-radial routes, more than 90 per cent of the objections to the general land use proposals concerned transport, and particularly the so called 'motorway box' (Button and Gillingwater, 1976). What puzzled the inquiry panel were three things: first, the planners' obsession with road building *per se* with little regard to the consequences for either public transport or environmental effects; second, the lack of any detailed appraisal of the tangible benefits of the road proposals, let alone an estimate of the resource costs; and finally, the lack of any firm proposals to improve public transport.

A similar story can be told from North American experience (see Schaeffer and Sclar, 1975). Boston provides the classic example of a city which literally ground to a halt because of a massive traffic jam. In a response, more than 10,000 off-street parking places were built, and more highways were constructed, including the Massachusetts Turnpike — a toll road for the last eleven miles into central Boston, aimed at ensuring an under-utilised section of road even during peak times. In one attempt to complete a crucial section of the network, one interchange resulted in the demolition of 479 buildings and the displacement of 326 households. The result of this sort of solution was to jolt the affected communities into organising themselves against the developers — which, ironically, were public agencies charged with producing tangible results in the public interest. Urban motorway protest was born, and it characterised this evolution of transport policy until the middle 1970s. In Boston for example, and largely as a result of this kind of action, the Governor announced a new transport plan which was based primarily on public transport, wiping out all previous plans for motorway developments.

It was not just the threat of public protest nor the new environmentalism which was responsible for this moratorium. The second major cause was brought about largely by external factors — a belated recognition that population forecasts were wildly high, and the impact of the Yom Kippur war of 1973 on global oil prices. These two events, in combination, served to demolish part of the foundations which underpinned the Age of Administrative Planning, but not the whole edifice. Wittingly or otherwise the protest against urban motorway development began to usher in a view that the state could no longer be trusted to represent the public interest (Cockburn, 1977).

If these were the external events which eventually ushered in the Age of Contestability, then they need to be complemented by two recurrent issues which were internal to the evolution of transport policy itself. The first was the issue of efficiency, the second the issue of subsidy. These have always been central to transport policy, and in all probability will remain so in the foreseeable future. Under the administrative planning regime, efficiency was always linked to integration and co-ordination rather than the notion of allocative efficiency. Efficiency through co-ordination and integration was assumed to eliminate wasteful competition. It was never assumed that it would replace competition altogether, but that co-ordination would provide the administrative framework, the operational guidelines for a rational pricing system, and integration would produce harmony and consensus in administrative terms. These were central to the 1947 Transport Act and the debates which preceded it. But as Aldcroft (1975) notes,

> on paper the arguments looked impressive, though at the time little was done to ascertain what benefits would accrue . . . or just how coordination was to be realised . . . Despite the government's earlier emphasis on the economic merits of unified ownership and control the proceedings were heavily concentrated on minor points and the terms of compensation to the owners. Both sides neglected the control issues, namely questions relating to technical modernisation

and investment, the integration of services and the criteria for charging.

The second point of increasing debate was over the issue of subsidy and the role of unremunerative transport services. As we noted earlier, this was proving to be a perennial problem for the railways, and was beginning to be an issue in road passenger transport, as the British Traffic Commissioners were noting in the late 1930s. In 1947, the BTC was charged with the responsibility of, on the one hand, making charges which would allow it to break even, taking one year with another, and, on the other, providing an adequate and efficient level of service and maintaining a properly integrated system of transport. The Commission presupposed this to mean that those services whose revenues did not cover their costs could be retained on social grounds through deficit financing and direct subsidy.

Had the BTC complete monopoly power over all sections of transport, including road space and vehicle use, it is possible to envisage a national pricing system which could accommodate these contradictory obligations. It did not however. Transport users were completely free to choose between alternative modes, particularly in road passenger transport. Moreover, competition had never been effectively ruled out from any transport policy statement, including the 1947 Act. The result meant not only averaging of charges and cross-subsidisation of services both within and between modes but, because of the nature of the changes affecting public transport, that over time the portion of service provision which was contributing surpluses began to diminish and the corresponding services requiring subsidy greatly increased.

Gradually since the late 1940s, the balance between administrative planning and permissible competition within a general regulatory regime had been changing in favour of the latter. Increasingly competitive enterprise has been inserted into public policy and legislation, beginning in Britain with the 1953 Act which took road haulage out of BTC control; then the 1962 Act which abolished the BTC, reorganised the railways yet again, and formally introduced competition between road and rail transport as a guiding policy; and subsequently with the 1968 Act with the advent of 'planned competition', in the form of separate state holding companies, between those commercially viable parts of different public transport modes, together with specific provision for subsidies to loss-making, socially desirable services.

The combined consequences of each of these relatively minor crises eventually led to the breakdown of the Age of Administrative Planning. By the late 1970s, a resurgence of neo-liberalist political philosophy became enmeshed in tangible and practical policy proposals which saw the regulatory and planning regimes of the previous fifty years as a major impediment to progress. The lead was taken in the United States with the appointment of market-orientated economists to key positions in the regulatory regime. The assault on regulation began before the passing of legislation, with the liberalisation of entry policy in road and air transport in the mid 1970s. By 1977 the majority of the Interstate Commerce Commission were economists and the chair at the Civil Aeronautics Board was taken by a notable liberal.

In Britain, the change is evidenced in the Department of Transport's published policy statement in 1977. This reaffirmed support for subsidies 'to meet social needs by securing a reasonable level of personal mobility, in particular by maintaining public transport for the many people who do not have the effective choice of travelling by car', but in practice led to a reduction in real terms in the amount of subsidy actually provided. The previously dominant and increasingly crisis-ridden regulatory regime under the Age of Administrative Planning was giving way to what appeared to be the emergence of a new era, the Age of Contestability.

The Age of Contestability

The key organising principles underpinning the Age of Contestability are deregulation and the reassertion of the application of allocative efficiency in transport policy. In common with neo-liberalist political philosophy, any barriers to free entry and exit from any market are seen as contrary to a redefined concept of the public interest. Instead of the state protecting the public interest through regulation and planned competition, the role of the state is to pursue the public interest through the process of creating conditions for efficient transport operations. According to this maxim, 'efficiency is best secured by giving the user maximum advice and allowing maximum competition' (Fowler, 1977).

This view gained the most currency in the UK when in 1981 the Court of Appeal found against the Greater London Council's policy of a 25 per cent fares reduction on London Transport operations. The key argument focused around the interpretation of the aims of the 1969 Transport (London) Act, which were to promote 'integrated, efficient and economic transport facilities', and clearly centred on public interest type arguments. One of the Appeal Judges interpreted this quite unequivocally: the Act did not include 'the question of social or philanthropic or political objectives'. The public interest was redefined further when the case went to the highest court in the land, the House of Lords, where the Law Lords ruled against the fares policy. Although the GLC had financial powers to support London Transport, no grants could be used for the purpose of social or transport policy, but to cover only any unavoidable deficit. Such an interpretation could never have been made using the public interest arguments which were characteristic of the Age of Administrative Planning, and which underpinned the spirit and purpose of the 1947 Act and its 1968 successor. Something in the mean time had changed dramatically, and that was to be found in the changing political philosophy which had come to dominate the 1980s. The decision in 1981 put the final seal on a process which had its roots in the increasing unease which characterised the period from the late 1960s, and which formed its first expression in transport policy in the Transport Act of 1980. Following a review of policy initiated in the late 1970s, and very heavily influenced by a new and equivalent mood emanating from the United States, this Act abandoned quantity controls of coach operations; dismantled price controls for all bus and coach operations; shifted the onus for new services away from the applicant having to justify a proposal 'in the public interest'; and set up trial areas whereby quantity controls did not exist at all.

In the United States, as noted earlier, *de facto* deregulation had effectively occurred in the late 1970s before legislation was enacted, beginning with civil air transport (Airline Deregulation Act, 1978) and followed by railways and road haulage (the Staggers Rail Act, 1980 and the Motor Carriers Act, 1980). Deregulation of the bus and coach industry was initiated in the

Bus Regulatory Reform Act, 1982. Although similar legislation has been enacted in Canada, only Australia has systematically followed the deregulation path thus far. It has been noticeably not followed with any enthusiasm in any European country. Whether this is testimony to the strength of the so called 'continental' philosophy, or whether there is a time lag built into the inevitability of its diffusion across the Channel, remains to be seen. Of one thing we can be sure: the complex issues and problems which we call 'the transport problem' will not go away, although they may change their shape and form, and neither will the contestability surrounding solutions to them through the application of transport policy.

References

Aldcroft, D. (1975), *British Transport Since 1914* (David and Charles, Newton Abbot).

Button, K. and D. Gillingwater (1976), *Case Studies in Regional Economics* (Heinemann, London).

Cockburn, C. (1977), *The Local State* (Photo Press, London).

Dunn, J. (1981), *Miles to Go: European and American Transport Policy*, (MIT Press, Cambridge, Mass.).

Fogel, R. (1964), *Railroads and American Economic Growth* (Johns Hopkins University Press, Baltimore).

Fowler, N. (1977), *The Right Track* (Conservative Political Centre, London).

Glaister, S. and C. Mulley (1983), *Public Control of the British Bus Industry* (Gower Press, Aldershot).

Gottmann, J. (1961), *Megalopolis* (MIT Press, Cambridge, Mass.).

Gourvish, T. (1980), *Railways and the British Economy, 1830–1914* (Macmillan, London).

Hall, P. (1974), *Urban and Regional Planning* (Penguin Books, Harmondsworth, Middx).

Hall, P. (1977), *The World Cities* (2nd edn), (Weidenfield & Nicolson, London).

Hart, T. (1983), 'Transport and Economic Development: The Historical Dimension' in K. J. Button and D. Gillingwater (eds), *Transport, Location and Spatial Policy* (Gower, Aldershot).

Hibbs, J. (1980), 'The Case for De-regulation' (mimeo, unpublished).

Hibbs, J. (1982), *Transport without Politics . . .?*, Hobart Paper 95 (Institute of Economic Affairs, London).

Plowden, W. (1973), *The Motor Car and Politics in Britain* (Penguin Books, Harmondsworth, Middx).

Pollard, S. (1982), *The Wasting of the British Economy* (Croom Helm, London).

Ponsonby, G. (1969), *Transport Policy: Coordination through Competition*, Hobart Paper 49 (Institute of Economic Affairs, London).

Schaeffer, K. H. and E. Sclar (1975), *Access for All* (Penguin, Harmondsworth, Middx).

Schnore, L. (1965), *The Urban Scene* (Free Press, New York).

Solomon, R. and A. Saltzman (1971), 'History of Transit and Innovative Systems', *USL TR–70–20* (MIT Press, Cambridge, Mass.).

Weiner, M. (1981), *English Culture and the Decline of the Industrial Spirit, 1850–1980* (Cambridge University Press, Cambridge).

3 URBAN TRANSPORT POLICY

Introduction

Policies for the regulation and control of urban transport have a long and varied history which certainly extends back beyond the days of the Roman Empire when laws were enacted to restrict the use of chariots in city streets. More recently we have witnessed quite dramatic swings in policy as urban policy-makers have tried to come to terms with the transport needs of modern, urbanised society. Many of these changes in policy have stemmed from the introduction of new technologies or have resulted from crises outside the transport sector but, equally, the perceptions and priorities of the policy-makers have themselves changed from time to time — often, but not exclusively, in response to social pressures.

The so-called 'urban transport problem' has been the focal point for policy formulation but this problem is itself far from a simple concept. Many of the early, post-Second World War studies of urban transport viewed the problem almost exclusively in terms of traffic congestion and, indeed, in terms of a very specific congestion problem — namely that relating to weekday, rush-hour, automobile congestion. As one of the first major US studies of the urban transport problem argued, 'Intracity freight movements and passenger trips at other times of the day or week can and do create important problems but these are almost of second order importance' (Meyer, Kain and Wohl, 1965). As we see in the following sections the urban transport problem is now viewed in a multidimensional context and is likely to continue to be viewed in this way in the future.

In examining the future of urban transport it is particularly important to study the actions of policy-makers over the past two or three decades. Not only have past policies left legacies of durable infrastructure and, in some countries, financial commitments but the experiences of previous policy implementation colour the way potential policy options are viewed. Equally, actions of previous generations of land-use planners have created a

41

wider urban environment which is going to limit and influence the courses of action open to transport policy-makers at least until the turn of the century. A later section of this chapter (pp. 52–9), therefore, examines the inherited legacy of the urban transport policy-maker.

The major external influences, those not directly or easily controlled by urban transport policy-makers, are discussed on pages 59–65 together with some assessment of the way these influences are likely to develop over the next decade or two. Clearly there is a degree of haziness over the dividing line between sections 3 and 4 (pp. 52 and 59); after all, previous policies with respect to urban transport have been based both on the circumstances of the time and upon the then projections of future trends. Equally, some of the seemingly external factors are themselves influenced by the action of urban policy-makers. Nevertheless, provided these potential links are borne in mind, the division between the future consequences of the past actions of planners and politicians and the impact of more general, socio-economic trends is a helpful one in assessing the future direction of policy.

The final sections draw upon the preceding sections to consider future developments in urban areas. Broadly, it proves necessary to divide urban areas into a number of different categories to do this. The nature of the problems confronting urban authorities in Third World countries is fundamentally different from the difficulties which are likely to be experienced in the West. Consequently, while one can discern certain common threads of policy, quite significant differences are likely to remain and, indeed, given the radically different circumstances, should remain in the actions which are likely to be taken. Pages 78–80, therefore, provide a few specific comments on urban transport policy in less developed countries.

The Urban Transport Problem

A useful starting point for a discussion of the urban transport problem is provided by the seven-element division devised by Thomson (1977) in his classic study, *Great Cities and Their Traffic* (See Figure 3.1). These seven elements, although differing in their detail and relative importance, are seen as common problems facing urban authorities throughout the world. They should not be viewed as independent of one another but rather as interacting elements of the overall urban transport problem. Delineation along Thomson's lines, however, does provide a useful discursive device and offers a framework for systematic discussion. The interdependences should not, however, be underplayed and, in fact, a lack of a full appreciation of their importance has resulted in cases of misguided policies being pursued in the past.

Figure 3.1: The Components of the Urban Transport Problem

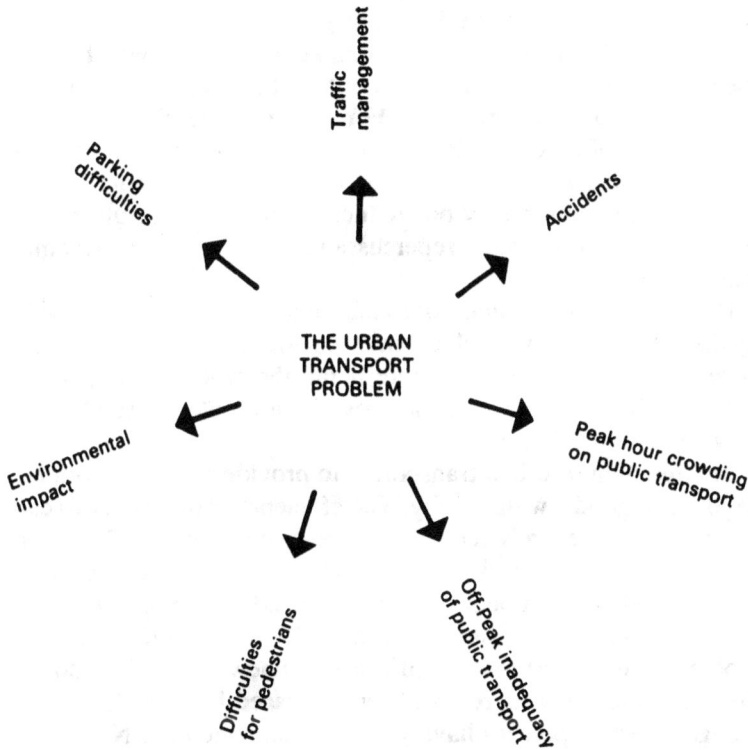

Source: Thomson, 1977.

The interdependences between the elements of the problem may affect policy formulation in two general ways.

Firstly, several of the problems generally coincide and, thus, policies directed at one problem area may act to alleviate other problems. In this way a series of policies may reinforce one another. Examples of this are not difficult to think of: traffic congestion, for example, normally results in high levels of pollution and noise and, therefore, policies to improve the flow of urban traffic can also enhance the quality of the environment.

Secondly, there are cases where policies, by tackling some problem elements, may actually exacerbate others. Once more, examples are easily come by and one can cite the increased difficulties experienced by pedestrians in cases where schemes have been initiated to speed up the flow of motor traffic.

The difficulty comes that while it is easy to generalise about the wider impact of a policy initiative, targeted on a given problem, in practice, local circumstances often make it uncertain whether the specific action will reinforce or counteract measures being taken with regard to other problems. For example, we cite above the general case of the coincidence of congestion and pollution; however, measures to speed up traffic by spreading it more widely across an urban area may not reduce pollution but result in detrimental environmental repercussions for formerly tranquil neighbourhoods.

Bearing in mind the limitations inherent in any attempt to divide up the urban transport problem into specific elements, it is helpful, nevertheless, to examine in more detail the types of element set out by Thomson and to consider how these problems are likely to be viewed in the future.

The function of urban transport is to provide mobility, both for people and goods, within a city. The efficiency with which this can be achieved is severely reduced if excessive congestion develops. A certain (optimal) level of congestion is desirable but it becomes excessive when additional vehicles so impede existing traffic that the *overall* benefits enjoyed from mobility begin to decline. The problem of excessive motor traffic is not a new one and economic theories concerning its removal can be traced back to the 1920s, though effective policies have yet to be implemented. Nor is it a problem confined to the Western world, for while it is true that Third World countries have much lower *per capita* car ownership levels (see Tables 3.1 and 3.2) their cities tend to be much larger and their roads less capacious. Further, the nature of the congestion problem has been changing somewhat in recent years.

A major difference between now and the immediate post-Second World War period is that the previously experienced pronounced peak in congestion during the rush hours is, in many

Table 3.1: Some Basic Transport Statistics from Selected Cities

City	Population in 1970 (million)	Per capita income (US$) (1970)	Autos per 1,000 people (1970)	Rate of auto growth (%) (1960–70)	Modal Split of Motorised		
					Auto	Bus	Other
Buenos Aires	8.4	1,800	73.9	12.1[a]	17	60	23[h]
Mexico City	8.6	1,275	78.3	10.5	19	65	16[c]
Singapore	2.1	1,100	73.0	6.7	24	43	33[g]
Hong Kong	3.4	850	26.2	7.1	22	55	23[d]
São Paulo	8.4	785	62.3	-	26	60	14[f]
Bogata	2.6	760	22.0	-	17	71	12[h]
Dar es Salaam	0.4	710	33.0	-	7	40	53[f]
Kinshasa	1.1	660	-	-	33	58	91
Kuala Lumpur	0.8	660	51.9	11.3	47	35	181
Lagos	1.4	555	22.8	15.5	12	-	88[d]
Bangkok	3.1	525	49.7	12.0	29	59	121
Seoul	5.5	440	6.3	22.0	8	89	31
Bombay	5.8	390	13.5	8.2	11	41	48[g]
Calcutta	7.4	270	13.0	7.2	8	34	58[e]
Madras	3.4	180	7.9	5.8	-[b]	54	46[i]
Tokyo	14.9	2,775	83.3	16.0	35	8	57[g]
London	10.5	2,550	222.0	5.2	59	24	17[i]
Paris	8.4	3,530	248.0	6.5	36	21	43[d]
Washington	0.9	5,390	316.0	2.0	68	32	i

Notes: a. For the country and not the city. b. Included in 'others'. c. Excluded trips by public motorcar. d. Data for 1965. e. Data for 1966. f. Data for 1967. g. Data for 1968. h. Data for 1969. i. Data for 1970. j. Data for 1971.

Source: World Bank (1975), 'The Task Ahead for Cities of the Developing World', World Bank Staff Paper No. 29; and W. Owen (1973), Automobiles and Cities, Strategies for Developing Countries (OECD, Paris).

Table 3.2: Persons per Motor Vehicle Registered

Region	1960	1970	1980
Africa	110	73	57
Asia	120	84	50
Europe	18.9	8	4.0
Oceania	4.5	3	2.5
North America	2.5	2	1.5
Latin America	57.5	28	15.3
World	23.5	14	11.0

Source: Maunder, 1983.

of the largest cities, gradually merging into a consistently slow traffic movement throughout the day. (This latter situation has long been established in many Third World countries where inadequate infrastructure has resulted in little temporal delineation of traffic speeds over the day.) A number of factors have contributed to this: changes in working practices (e.g. more 'flexi-time') and in the composition of the inner-city labour forces (i.e. relatively more technical and executive personnel at the expense of clerical workers has both increased the proportion of car-owning commuters and combined this with a wider spectrum of work-hours) have been influential but, equally, increases in traffic management control problems at off-peak times (e.g. with regard to parked vehicles) and in the amount of freight moved over urban roads have also contributed to the situation of uniformly slow traffic movement throughout the day.

In one sense, however, congestion is self-regulating and there appears to be a mechanism which prevents excessive congestion becoming totally debilitating to the vitality of any city. As congestion slows traffic so some people and firms move out of the city, often to suburban locations, or modify their activity patterns so that busy transport periods are avoided. (If we look, for example, at recent trends experienced by many cities towards greater decentralisation then there is some evidence that high levels of inner-area congestion has been one of the contributing factors.) Congestion is, therefore, contained in the absence of effective public policies at some threshold level. What this threshold will be is likely to vary between cities but there is evidence of a remarkable constancy of rush-hour speeds of around 16 kmph. to be found across a wide range of major cities (Thomson, 1977). It is perhaps worth noting that this compares with an optimal speed of about 25 kmph. which several desk-top studies suggest policy-makers should aim for in London. In some cities local policies of public transport provision and extensive road-building pro-grammes have pushed up the peak-period speed but evidence from the USA (e.g. Meyer and Gomez-Ibanez, 1981) indicates that this is only a short-term deviation from the unrestrained equilibrium.

Clearly questions of parking provision are closely related to those of traffic congestion. In larger cities the proportion of through-traffic is much less than in smaller centres both in terms of passenger vehicles and goods vehicles (see, e.g. Reynolds, 1960) and this poses problems of catering for those requiring parking space. In most cities there is a shortage of parking although in many cases this now stems from a deliberate policy designed to encourage people to use public transport modes. Parking prob-lems are generally perceived to be most severe in the USA partly because of the high level of car ownership but also because of the relatively poor quality of public transport. Recent years have witnessed two major changes in the dimensions of the problem, especially in the UK. Firstly, areas of land laid bare in uncompleted redevelopment programmes and subsequently taken into use as parking space are now being developed and, thus, reducing the supply of informal parking available. Secondly, there is mounting evidence that parking policies, which have traditionally been thought of as generally acceptable public controls, are consistently being violated (see e.g. O'Neil, 1977 and Elliott and Wright, 1982).

Public transport in urban areas has come under increasing scrutiny since Thomson's study. While there is still concern about the capacity of many systems to cope with peak-period demands placed on them and the quality of service offered at off-peak times, especially at weekends and evenings, the overall question of efficiency in provision has become a topic of debate. In particular, there is the mounting problem of financing unprofitable services. The extent to which systems in the developed countries receive direct operating subsidies, while a universal fact, does still, how-ever, differ quite markedly (see both Figure 3.2 and Table 3.3) as some countries have attempted to keep real costs of fares constant in the face of rising costs (e.g. as in the case of the UK) while others have been willing to pay higher subsidies to keep the nominal levels of fares constant (e.g. most US cities). As can be seen from Figure 3.3, the divergence between fare base revenue and costs in the latter cases has grown quite noticeably in recent years and while the actual levels of subsidy differ between countries the explosive nature of the situation is uniform. Additional to the problem of direct subsidies is the extent to which cross subsidies are posing serious distortions in the markets for urban public transport. There is now substantial international evidence (e.g. Wachs, 1981 and Monopolies and Mergers Commission, 1982) of considerable cross-subsidisation within the urban

Figure 3.2: Trends in Subsidies in Industrialised Countries

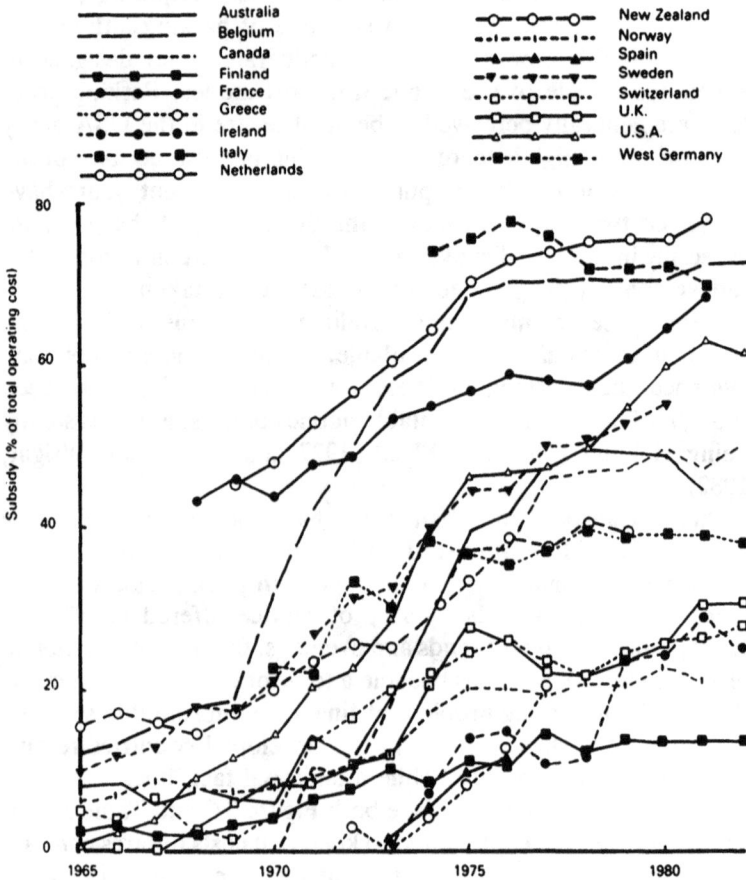

Source: F. W. Webster, paper presented to the ECMT 10th International Symposium on the Theory and Practice in Transport Economics held in Berlin, 1985.

Table 3.3: Comparison of Levels of Urban Transport
Subsidy. Subsidies as a percentage of costs for 1977

Country	Small Towns (%) (Population < 400,000)	Large Towns (%) (Population > 400,000)
Australia	48	29
Canada	37	26
France	26	55
Netherlands	57	67
Sweden	25	41
United Kingdom	14	21
United States	24	26

Source: Reid, 1983.

Figure 3.3: Operating Costs and Revenue (in money terms) for US Urban
Transit 1950–80

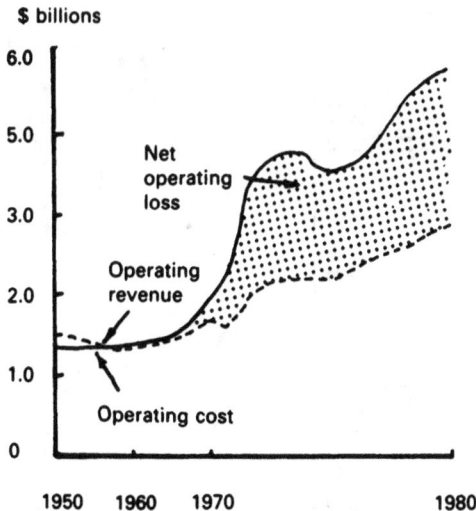

Source: Fielding, 1983a.

transport sector by time of day, length of trip and by route. Not
only is this raising questions of equity and managerial efficiency
(i.e. is a specific service cost-efficient?) but in the longer term,
given the different elasticities of demand for various services, it is
likely to worsen the financial situation confronting public transport
undertakings (see Fielding, 1983b).

The question of off-peak provision of urban transport services

Table 3.4: Changes in the Distribution of Populations by City Size 1950–70

		Percentage of population living in cities of (thousands)				Percentage of population in cities of 100,000 + inhabitants
		<100	100–199	200–499	500–999	>1,000	
Less developed Countries	1950	48.95	8.89	12.18	10.45	19.52	16.46
	1970	42.51	7.59	11.83	8.70	29.36	25.68
More developed Countries	1950	42.74	8.62	11.84	9.25	27.55	51.62
	1970	36.56	8.88	12.74	9.83	32.00	63.97

has also taken a somewhat different form in recent years. The spread which has taken place in urban residential land use together with the decentralisation of many recreational and retailing activities means that conventional radial transport systems are no longer necessarily appropriate. To reach out of town shopping, medical or leisure centres it is often necessary to make cross-town trips. The people generally needing (or wishing) to use these facilities are more often than not those with the least access to private transport.

Public transport problems in less developed countries are somewhat different. In Third World countries the capacity of public transport systems for work trips is often limited relative to the demands placed on them (see again Table 3.1). The low — albeit growing — levels of car ownership in most Third World countries means that considerable reliance is placed on public transport modes for journeys to work but there are only limited funds for investment in circumstances where populations have insufficient incomes to pay economic fares. The vast explosion of populations in Third World countries combined with increased levels of urbanisation (Table 3.4) have also put considerable strains on the existing public transport systems which may well have coped in the past.

Injury and death are emotive issues and, thus, safety is a key element in urban transport policy. The direct consequences of urban transport accidents are well documented and the issues have essentially remained unchanged in recent years. As with congestion, there seems to be a threshold. People are willing to accept a certain level of perceived risk of accident (which may differ from actual accident rates) but, beyond this, public pressure causes the authorities to act. In most developed countries there is some indication that this threshold has fallen in recent years with growing affluence but the reverse seems the case in Third World countries where incomes are low and employment may depend upon making dangerous journeys (e.g. on the outside of a bus or tram). The situation is not helped by the poor state of roads, the undisciplined behaviour of road users and inadequate traffic management (see Thomson, 1983).

One aspect of safety which is changing relates to the carriage of dangerous consignments of chemicals and other substances. The potential harm that spillage or leakage could inflict on an urban population has become a topic of public concern following a series of accidents (e.g. spillages of chemicals in Canada in 1982) and policy-makers are being forced to direct their attention to the carriage of dangerous substances.

Transport (Thomson, 1974) has been described as an engineering industry carried on outside the factory. In return for the mobility it affords, motorised urban transport imposes consider-able environmental costs on the community in terms of the pollution it generates, the noise it creates and the visual intrusion it imposes (for a plethora of numerical detail see Sharp and Jennings, 1976). There is always a problem in actually defining the environmental dimensions of the urban transport problem, in part because many of its components are simply difficult to quantify but there are also difficulties with perception (e.g. people tend to dislike diesel fumes more than those given off by petrol engines although the latter are more likely to damage the health) and inadequacies in our knowledge about which pollutants are really harmful and in what doses. In general, questions of environmental damage arise more often in wealthier societies where attention has progressed from ensuring the provision of necessities to concern about the 'quality of life'.

The final element of the urban transport problem isolated by Thomson concerns pedestrian movement. Walking is the most important single mode of transport in cities but, by its nature, it interacts with mechanised forms of transport — in the USA about 40 per cent of urban traffic accidents involve a pedestrian or cyclist. Striking the appropriate balance is not easy but there is a tendency for transport planners to focus on the needs of drivers and public transport to the neglect of pedestrians. There is a further dimension to the problem. Many non-car owners and those who cannot afford public transport services are often dependent on walking to move around the city and even those living in a car-owning household often have no access to it. Recent trends towards increased suburbanisation place such people at a severe disadvantage in terms of the level of mobility they now enjoy.

Post-Second World War Urban Transport Policy

Given the multidimensional nature of the urban transport problem it is not surprising that post-war policy has experienced periodic shifts of emphasis and approach (for a more general view see

Chapter 2). As a broad generalisation, up until the 1920s urban policy had tended to be reactive and to be strongly influenced by major technological changes (Schaeffer and Sclar, 1975). Urban areas were limited in size and restricted in potential land-use configurations by the inadequacies of transport technology. While the railways offered improved inter-urban transport, urban activities remained centred around a rather limited core. Advances from the mid-nineteenth century in passenger transport (first the horse-drawn tram but later electric trams, subways and elevated railways) permitted some outward movement but industry and retailing were restricted by the limitations of horse-drawn drays. The motor lorry, which became widely available after 1918, removed the geographical limitations on industrial location and permitted urban areas both to expand in size and to develop suburban centres.

In the pre-1920s environment, policy was directed mainly at encouraging public transport provision with enabling legislation and, on occasions, financial aid. Even at an early stage, however, there was some concern about environmental matters and it is no accident that, where there exists a prevailing wind, railway yards tend to be located on the leeward side of a city. The fact that steam traction was only slowly introduced into urban areas — in elevated systems in US cities and underground in the UK — while partly explained by the inefficiency of steam locomotion for stop-go operations also stems from concern about the soot and smoke generated.

The 1920s and 1930s had seen some more direct involvement in policy-making, especially with regard to the provision of public transport services. Concern grew over safety and in most countries various forms of controls were initiated, regulating the quality of vehicles and the conditions of operations permitted when offering public transport services.

While conceptually different, the question of social service provision became entwined with quality control debates of this period. A substantial part of many urban public transport systems had only attracted private investment because of the enhanced land values that improved access brought about. These were once-for-all gains, however, and, given only the limited interest in charging economic fares that higher land values engendered, when reinvestment and replacement became necessary insufficient private finance was forthcoming. To retain public transport services and, as a *quid pro quo* for accepting fare restraint at a time when investments needed funding, local authorities provided protection for public transport undertakings against competition from alternative modes. Regulations and licensing systems were introduced which essentially forced jitneys and other forms of paratransit from the market and, in the case of urban bus operations, offered protection from new entrants. This provided a framework whereby unprofitable operations were financed from the monopoly profits earned on other, protected services.

While the immediate post-Second World War period saw the peaking in the demand for public transport, rising affluence and increasing levels of car ownership very rapidly began a switch towards private modes. To cope with the new situation major urban road-building programmes were initiated in the USA. European states soon followed a similar strategy especially in cities devastated by military action. The focus was on improved access but problems soon emerged. These problems differed between countries and varied in their intensity. Similarly, the policies adopted had specific national pecularities and their implementation differed in its timing and commitment but the general picture is of universal applicability. (A few specific words on Third World countries are offered later in the section.)

Firstly, by the mid-1950s there was a developing view that the impact of urban road construction was on land-use patterns rather than easing movement within cities. Essentially urban sprawl was seen to be occurring as people were moving to suburbs rather than using the possible time savings for non-travel purposes (US one-way commuter trips still averaged about 20–25 minutes despite much faster road speeds — see e.g. Wingo, 1961). Whether this effect is socially desirable is debatable. It led to a greater polarisation of society with higher, car-owning households moving further from the urban core. Additionally, as time passed, Down's Law began to bite (*viz.* 'on urban commuter expressways, peak hour traffic congestion rises to meet maximum capacity', Downs, 1978) and brought forth pressure for more road building.

Secondly, concern began to grow around the environmental impact road building was imposing, especially on non-travellers. This was later to reach a peak in the UK with the public protests against the construction of 'Westway' in London and in the decision in the USA not to complete the Embarrdero Freeway in San Francisco. Initially the solution was perceived in improved road network design with less emphasis on pure traffic engineering efficiency and more on wider social and land use criteria (e.g. Ministry of Transport, 1963). The rather idealistic approach advocated involved separating transport from environmentally sensitive activities. This, however, could only provide a very long-term solution to the emerging situation and could only do this with an extremely high resource commitment.

From the early 1960s a general disillusionment with the road-building 'panacea' began to emerge and attention switched to the revitalisation of public transport modes. These, and especially fixed track systems, appeared to offer an alternative solution to the access–environmental conflict by providing a means of mass movement during peak periods on track separated from residential land. Electric traction was seen as quieter and cleaner than petrol-driven automobiles. The absolute decline in the use made of public transport modes and the lack of adequate investment over the preceding forty years were to be reversed. In the USA and many European countries capital was injected into the construction of new systems and the updating of old ones. Organisational reforms — e.g. the 1968 Transport Act in the UK and the Urban Mass Transportation Act of 1964 in the USA — put urban public transport undertakings on a new financial footing with large units (increasingly under municipal ownership) being formed and integrated more closely with other dimensions of urban policy-making. Central government funds were made available in most countries for new investments. Novel forms of fund-raising were initiated with, for example, a special payroll tax being levied win France to raise revenue for public transport investments, a percentage of the West German mineral oil tax being earmarked for public transport subsidies and a special fund being established in Sweden to assist public transport.

The new and improved public transport systems, however, could not attract sufficient patronage at cost-recovery fare levels to alter mode splits significantly (indeed, in many cases no viable fare-ridership/service-level combination emerged to permit cost recovery). Since road users were still essentially unrestrained in the use they made of their cars (planning policy in London, for example, was still encouraging the provision of adequate parking facilities in all new commercial buildings) and not penalised for the congestion costs they generated, the incentive to switch to public transport modes, even at very low fares, was extremely small. The carrot was proving ineffective without the rod. Further, motorists only perceive the immediate costs of making a trip (normally the time involved and possibly some idea of the petrol consumed) but take no account of depreciation of the vehicle, wear on tyres etc. in deciding whether the trip is justified. In contrast, public transport fares, if costs are to be recovered, must include allowance for the *full* costs of a journey. Questions have also subsequently been raised about the efficiency of public transport undertakings both in terms of the objectives pursued by management (see e.g. Cooter and Topakiam's 1980 study of the BART system in the USA) and the levels of cost incurred (e.g. see Ortner and Wachs, 1979).

The short-term result was a rapid rise in the level of operating cost subsidisation — total direct subsidies to US urban transit, for instance, rose tenfold between 1970 and 1978 to reach $5.2 billion. Short-term justifications rested on expediency surrounding the 'oil crisis' while longer-term social arguments, especially in relation to the revitalisation of inner-city areas, were also advanced in support of subsidy policies.

By the late 1970s a number of factors led to a gradual reversal in thinking regarding the public transport orientation of policy. The sheer size of subsidy finance was in itself a problem especially at a time when macro-economic policy in many countries was concerning itself with the impact of public borrowing needs. Further, the earlier studies of the relationship between subsidies and public transport costs were beginning to appear (see above pp. 47–51). Both of these developments may have been accepted by the policy-makers if the objectives of subsidy policy were being achieved. While the improved provision of transport services was helping improve the mobility of non-car owners it was by no means certain that this was being achieved with an acceptable level of cost efficiency. Further, public transport was not making a significant impact on environmental problems or congestion. While some short-term improvements did emerge, these did not prove long lived in a situation of unrestrained car use. Equally, the impact of public transport on the economic revival of inner areas was deemed negligible although here it is possibly fair to say that the time span for a genuine evaluation was too short.

The emphasis of the 1980s has switched towards greater efficiency in the provision of public transport and the most effective use of subsidies combined with giving more weight to policies designed to restrain the use of the private motor car.

Efficiency is difficult to measure in the context of urban public transport but most major systems have now devised criteria to provide guidelines in their specific circumstances. London Transport, for example, reviews its system and fares structure in terms of getting the maximum passenger-miles from its subsidy. It is also noticeable that most Canadian undertakings have, since the mid-1970s, related the costs of their labour and capital inputs to the revenue derived while the Dutch have a series of service indices reflecting user requirements *vis-à-vis* cost. The question of ownership and organisation has become central to the policy debate in both the USA and the UK (see e.g. Department of Transport, 1984). Increasing empirical evidence (e.g. see Anderson (1983) study of US data) suggests that municipal ownership results in higher costs and this has led to mounting pressure in some quarters for the privatisation of public transit undertakings and the provision of contract services. Coupled with this has been a questioning of the whole concept of traditional urban public transport. The proven high costs of peak-period provision, for example, suggest that easing access to this market could reduce the levels of subsidy support required, while the geography of expanded areas indicates that more personalised, paratransit modes may offer a more effective public mode to conventional, rigidly timetabled mass transit forms. It is no accident, therefore, that jitneys appear to be making a return, both in the UK and in the USA, albeit in rather different forms (e.g. as shared taxicabs) to those found in many Third World countries or in the West during the first twenty years of this century.

This brief policy review has concerned itself with general trends in economically advanced countries. The situation in less developed countries has been somewhat different. Few of these countries have the resources required for large-scale investment in modern transport infrastructure and have (since colonial days in most instances) been reliant upon overseas aid to carry out their urban investment programmes. Further, a lack of expertise in the majority of these countries has meant that considerable policy advice has been sought from 'experts' from Europe and the USA. This has had the impact of importing overseas thinking and ideas which, retrospectively, may not have resulted in the most appropriate policies being pursued given local circumstances and needs. In particular, transport in most cities of the Third World involves a much wider range of modes than in advanced economic society and familiarity of these was not a feature of many of the consultants' experience. Against this, however, must be set the somewhat different organisational structures and institutional frameworks which have made the introduction of certain policies less difficult than in many developed countries, thus offering more scope for experimentation.

Up until the late 1950s, low car-ownership levels coupled with a heavy reliance on non-mechanical transport meant that the urban transport problem as we now define it was limited to a small number of more advanced Third World cities. From the late 1950s, however, rising numbers of automobiles and an accelerated pace of urbanisation necessitated some action (see Thomson, 1983). The consultants called in, especially by international agencies, attempted to adapt the long-term comprehensive, land-use/transportation planning methodology then in vogue in the USA and, to a lesser extent, in Britain to local needs. While the diagnostic phase of such studies threw up useful information, the policy approaches which tended to accompany the approach involved heavy reliance on large-scale investments, especially in roads — a solution more suited to Western economies for which, indeed, the land-use/transportation methodology was designed. Even if the resources could be found for the networks which were advocated they were in themselves seldom helpful to urban economies with populations expanding at rates outside the range of the computerised projections and which were dependent upon industries often run on commercial lines rather than involving capital-intensive factory production. Further, the approach offered only long-term solutions to problems which were immediate and becoming critical. In terms of equity, road-orientated schemes also tend to be highly regressive in most less developed countries where the majority of urban journeys are made on foot or by bicycle.

From about 1974 the approach changed and the emphasis switched to piecemeal, incremental improvements which would act almost as fire-fighting measures. The oil crisis and worldwide recession were the immediate causes of the change in focus but more general trends were forcing policy changes of this kind in any event. Traffic management and low-cost investments were initiated to make better use of existing infrastructure, and public transport, particularly bus services, was enhanced by improved repair facilities and the introduction of new, more fuel-efficient vehicles. An emphasis was placed on repair and maintenance, mainly to assist in longer-term transport supply but also to improve the quality of existing transport provision.

The advantages of the approach are clear — there are immediate improvements in the quality and capacity of the available transport system. The changes introduced can be specifically geared to local needs rather than designed to meet a more general blueprint. Further, the resource costs involved are likely to be relatively small, leaving funds available for allocation to other sectors. By being spread across many projects they would also appear to benefit more people. What are less immediately obvious are the potential weaknesses. The piecemeal nature of the approach means that essential co-ordination may be sacrificed, leading to gradual, long-term inefficiencies emerging. Equally, the short-term nature of the underlying philosophy means that some projects which involve an extensive gestation period but offer a very high, long-term social return are sacrificed for the sake of short-term reactions to immediate problems. These latter weaknesses of the piecemeal approach are likely to result in mounting problems in the future — a concern which is now being expressed by several of the international development agencies. We return to look at the specific policy implications for the cities of the Third World in section 6 of this chapter.

External Trends

Transport policy must respond to changing urban conditions. Unfortunately for policy-makers, recent years have seen significant shifts in trends (which have in the past been treated as fairly stable) while in other instances there have been appreciable accelerations in the pace of change which had previously been viewed as relatively constant and even. While some of these trends are to do with shifts in the location of economic activities others reflect changes in tastes, factor costs and political attitudes (see Fothergill and Gudgin, 1982). Certain of the economic trends are long-term and can, after some careful study, be predicted with some degree of certainty but this is less so with political and taste considerations.

Perhaps the most pronounced trend in Western cities is that of spatial decentralisation of productive plant in the manufacturing sector. As can be seen from Table 3.5 the expansion of employment in the inner areas of the UK in the immediate post-war period was completely reversed in the 1960s and early 1970s as industry expanded in suburban locations. This is certainly not a uniquely UK phenomenon although there are national differences. In the USA, where decentralisation has perhaps been most pronounced, some 542,000 jobs were lost in New York, for instance, between 1969 and 1976 while in total something like 1.7 million jobs were lost to the older manufacturing centres located in the northern industrial areas. In mainland Europe, the trends have been less severe but there has been a marked decentralisation in the larger urban areas of West Germany, Holland, Switzerland and Austria. Figure 3.4, for example, depicting employment trends in Brussels provides a fairly typical picture of what has happened in the established manufacturing countries of Europe. The southern European states seem to be drifting in a similar direction although the current is both somewhat calmer and slower. A similar picture is emerging with regard to Japan.

Table 3.5: Intra-Urban Population and Employment Trends in Great Britain

	1951–61		1961–71	
Population	*000s*	*%*	*000s*	*%*
Urban core	486	1.9	−729	−2.8
Metropolitan rings	1,721	13.3	2,512	17.2
Outer metropolitan rings	245	3.1	786	9.8
Unclassified areas	−21	−0.9	−32	−1.4
Employment				
Urban core	902	6.7	−439	−3.1
Metropolitan rings	293	6.6	707	15.0
Outer metropolitan rings	−14	−0.4	130	3.9
Unclassified areas	−56	−5.5	−7	−0.7

Source: Young and Mills, 1983.

Changes in the nature of production (a topic also discussed in Chapter 6, see pp. 159–68) means that modern industry is more mobile and tends to be more land-intensive in its production processes. Essentially, the land–output ratio has risen to facilitate conveyor belt production — the resultant rise in capital and labour productivity more than compensating for the higher land

Figure 3.4: Trends in Employment in the Built-up Area of Brussels

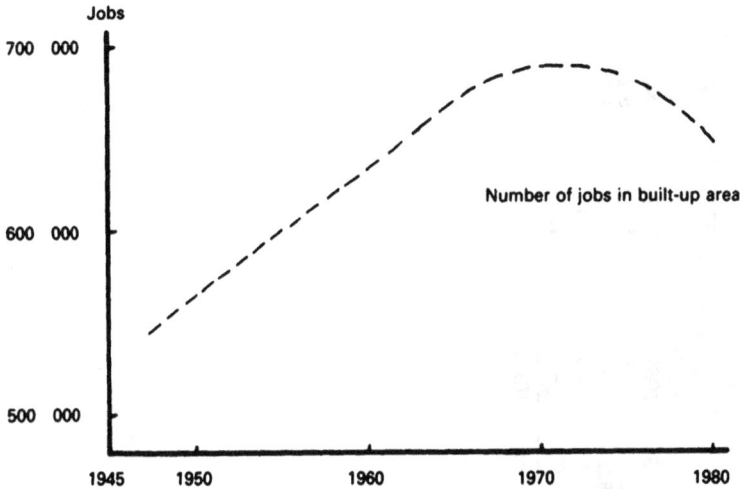

Source: M. Gochet (1985), 'Passenger Transport's Investment and Financing Decisions in Urban Transport' in ECMT, *The Evaluation of Past and Future Transport Policy Measures; 10th International Symposium on Theory and Practice in Transport Economics* (ECMT, Berlin).

costs. Additionally, many of the newer industries serve larger geographical markets but produce less bulky commodities, a combination of factors which lead to locations near high quality transport facilities and away from old, established city centres. Superimposed on this were the policies of many urban authorities which until the mid-1970s actively encouraged outward movements of industry for essentially environmental reasons. As these forces develop and land-use patterns of a more general kind adjust, this is likely to require a reorientation of transport policy with a greater emphasis on suburban policies.

Coupled with and overlapping the decentralisation of industry there has also been a general trend towards deindustrialisation — a swing from manufacturing to the service sectors — in most developed economies (see Table 3.6 for some indications of the effects of this on UK cities). The explosion in information technology coupled with the increasing demands for services which accompany higher incomes is likely to sustain this growth for some time to come. Figure 3.5 provides an example offering data on trends and projections of the US employment situation. The

Table 3.6: Percentage Change in Employment in the Inner Areas of the UK

	Manufacturing		Construction		Services		Total employment	
	1969–71	1971–3	1969–71	1971–3	1969–71	1971–3	1969–71	1971–3
Merseyside	−16.6	−7.0	−23.3	−5.9	−9.1	−1.2	−9.2	−2.7
South-east Lancashire	−9.9	−9.5	−21.1	+8.8	−4.6	+0.8	−6.9	−2.1
Tyneside	−4.7	−4.5	−20.4	+15.6	−4.3	−0.3	−5.0	−0.3
West Midlands	−7.1	−7.4	−11.0	+5.0	−2.8	+1.7	−3.9	−3.0

Source: *Study of the Inner Areas of Conurbations, Department of Environment (DOE) June 1975.*

implication of this shift towards the service sector is initially likely to be increased pressure on transport systems — especially road networks — providing access to central urban areas where service sector employment is located. This is a trend reinforced by the productive decentralisation of manufacturing where the internal service departments of firms (e.g. marketing, finance, etc.) tend to remain in central urban areas to permit interaction with external service activities. In the longer term, however, both advances in information technology (which will reduce the necessity of the face-to-face contacts normally used to justify the physical concentration of service sector activities) and the natural tendency for many services eventually to follow manufacturing will

Figure 3.5: US Employment by Sector. In millions of people on private payrolls — seasonally adjusted

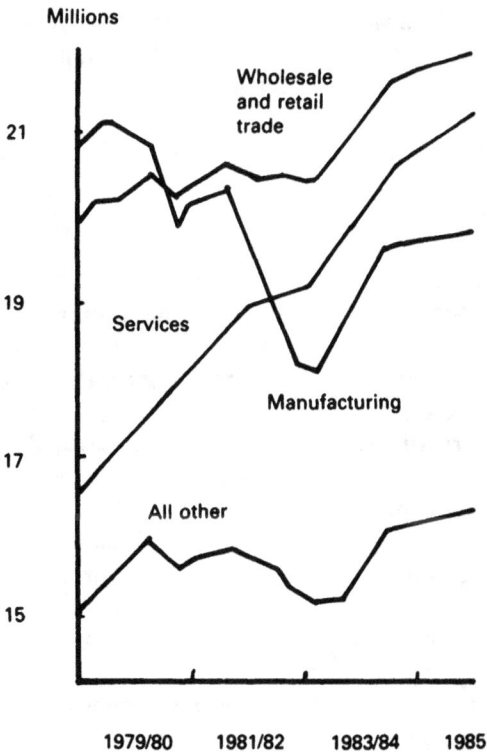

Source: Fortune, 1984.

considerably reduce this pressure. The greater flexibility offered by new technology will also offer scope for a spreading of peak travel demand (and should also assist in the acceptance of some of the restraint policies discussed in the following section).

Evolution in the structure of industry are also resulting in greater use being made of part-time and, especially, female labour in many Western economies. In part this may be explained by the lower wages which it is often possible to attract such labour with but also much of the male labour displaced by the changes discussed above has been slow to accept employment in the types of industry where growth has occurred. More important, however, would seem to be the greater activity rate now found amongst women. In the longer term this trend is likely to be curtailed both as the rise in female activity patterns reaches its zenith and as the male labour force adjusts to the new situation. But in the shorter term, since females tend to have more limited access to private transport the implication is that there is likely to be a continuing need for public modes to be available although, given the less than standard hours worked, especially by part-time labour, this may not be of the traditional form.

Clearly residential patterns are closely linked to developments in employment opportunities. As we have seen (see again Table 3.5), recent tendencies have been for people to move away from inner-city areas. This exodus has, however, not been spread evenly across the population but has tended to be concentrated among the young skilled and professional groups. In the USA, it has been further segmented by the relative immobility of black workers. The result is that the remaining residents of inner areas tend to experience low incomes and high rates of unemployment coupled with relatively high numbers of elderly and children. The transport requirements of these groups clearly deviate from those of traditional commuters.

There seems no reason to anticipate a dramatic reversal of this trend although an inevitable slowing-down process will occur as more people achieve their ambitions. More generally, however, the changing nature of industry is likely to produce a continual shift of population between cities with moves from urban centres in depressed regions to those offering longer-term employment prospects and more desirable living conditions (e.g. the 'sunbelt effect' in the USA). These longer distance migration trends will necessitate expansions in infrastructure and investments in transport facilities in the recipient regions while calling for policies to ensure appropriate, evolutionary adjustments in contracting urban settlements.

The Future Direction of Urban Transport Policy

As we saw at the beginning of this chapter there are many elements to the urban transport problem most of which transcend the purely urban dimension to form part of many transport concerns. For expositional reasons, and to keep things manageable, in this section we concentrate on the probable directions of future policy in three key areas — traffic restraint, public transport provision and urban transport infrastructure. This division is not the only one possible but it does provide a basis for discussing the main dimensions in which policy is likely to exert any form of impact. Policy in these areas also influences and is closely tied in with such key issues as fuel conservation, environmental protection, social-service provision and urban land development.

Obviously the division we have selected ignores many overlaps in policy formulation and does not do full justice to the important interactions between decisions on vehicle restraint, public transport operations and infrastructure but it does correspond to general administrative arrangements (which seem almost enshrined in tradition). Also it permits policy discussion to be examined in a systematic fashion. The section as a whole concentrates mainly on urban transport in the advanced Western economies, but, as we see, there may be things to be learned from the experiences of Third World nations.

Traffic Restraint Policy

There seems little doubt that car ownership will continue to grow throughout the world over the foreseeable future. The comfort, convenience and privacy offered by the mode is unlikely to be challenged by potential alternatives. Accommodating the increased car traffic implied by this trend is, in the absence of any restraint policy, only possible if massive investments are made in new urban infrastructure. It is, however, both unlikely that resources will be available for such an expansion and questionable if they should be committed to urban road construction if they were. There is an opportunity cost associated with road building which means that expansion of the system diverts resources from being employed elsewhere in what could well be, from an economic or social perspective, more important sectors. It is unlikely, for instance, that many people would want the scale of road provision necessary to accommodate the unrestrained growth in car traffic given the sacrifice of hospitals, parks, buildings etc. that it would entail. The expansion of urban road networks, as we have pointed out earlier, is also going to generate additional traffic by its very existence, meaning that road-building would have to continue just to keep pace with the expanded traffic volumes it is encouraging — Down's Law again.

In these circumstances enhanced measures of traffic restraint are likely to play a significant role in future urban transport policy. If one examines recent policy developments there is ample evidence that policy is already moving in this direction although it seems probable that both the pace of movement will accelerate and the nature of the restraint measures will differ from those which have been employed to date.

Restraining the free use of roads poses serious political difficulties in most Western democracies; it essentially goes against long-standing ideas of free movement ('the freedom of the road'). In practical terms, it is also often not easy to design policies which are cheap to implement and do not have serious side effects which can, for example, lead to economically undesirable trends in urban land-use patterns. For these reasons most of the policies which have been favoured to date have been adopted as much for their political acceptability and administrative convenience as for their performance as tools of traffic restraint. This is not, of course, to say that some direct measures of traffic regulation have not been quite readily accepted — e.g. priorities at junctions, driving on one side of the road, compulsory driving tests etc. — but these are easily presented as simple 'quality controls' necessary for obvious reasons of safety. Indeed, in most people's eyes they are not even viewed as traffic restraints at all.

Physical traffic management policies of a more general kind (e.g. one-way streets, traffic signals etc.) have gradually been accepted as necessary tools to ensure the smoothest possible flow of traffic in an urban area and undoubtedly their use will be refined as time goes on. They act as a restraint in the sense that they say how a road system may be used but they are not a complete restraining measure in that they do not actually regulate the *total* volume of traffic. Another way of looking at them is that they maximise the potential economic capacity of a road system but do nothing to ensure that the traffic volume is limited to that capacity. Unlike simple quality-type controls traffic management schemes often meet considerable local objections because they frequently involve funnelling traffic with many individual trips actually being longer than without them even though the overall flow is larger.

The use of parking restrictions and pricing have been, for the past twenty years or so, the main instrument of pure restraint in most Western cities. (Although in many cases the actual charges deviate significantly from the optimal level.) Their limitations are obvious, they do not reduce through traffic, and may even encourage it, but nevertheless politically they initially proved relatively easy to introduce. Many of the fears of planners that excessive reliance on such policies may result in undesirable land-use distortions and, especially, add to the forces causing suburbanisation, have proved well-founded. Essentially parking fees are insensitive to journey length and thus impose relatively less of a burden on longer distance commuting. It is now becoming clear that much greater care must be used in the employment of such policies and, indeed, with the actual provision of parking spaces. Recent research findings in the USA, for example, into methods of estimating optimal parking fees combined with limited relaxations on parking controls in cities such as Washington (usually to stimulate downtown retailing activities) indicates that a degree of rethinking is already taking place. The need in the future is to redefine parking policy so that it co-ordinates more closely with other policies designed specifically to restrain moving traffic. Essentially, parking and driving are complementary and, as such, require appropriate, co-ordinated and consistent policies to be applied to both simultaneously.

Given the gradual realisation that parking policy should ideally be directed specifically at the allocation of parking spaces and is not, except in clearly defined circumstances, very effective as a surrogate instrument aimed at other objectives, there is a need to initiate new measures aimed directly at traffic congestion. Direct rationing of road space is one possibility but administratively this is likely to prove cumbersome and politically it is inconsistent with the ideas of market processes which traditionally dominate thinking in most Western economies. Further, given the tendency for political power eventually to dominate physical allocative systems, irrespective of any original intentions of allocating by need or efficient use, it is not altogether certain that direct rationing by, say, a permit system is likely to serve the best interests of the urban community. The misuse by some people of parking permits for the disabled in the UK over recent years provides a practical illustration of the difficulties which can accompany direct allocation procedures.

An alternative approach which has to date been adopted in one country and experimented with in another is to apply the price mechanism directly to the urban road system and to sell road space in the same way as other commodities are sold. Just as there is a price which acts as a mechanism to equate the supply and demand of, say, television sets and prevents queues from forming up to purchase them at artificially low prices, so it is possible to estimate a price for road space which rations it out amongst those desiring to use it. (A rigorous account of the relevant theory is contained in Morrison, 1986.)

In the past a vast number of desk-top simulations have been performed to assess the potential of this economic approach. A form of area licensing, for example, was seriously considered for parts of London in the late 1970s while the US government has tried unsuccessfully to tempt urban authorities to experiment with road pricing by offering financial and technical assistance. In Australia, the city authorities of Perth and Adelaide looked at the idea only to back away in the face of public objections while Kuala Lumpur went as far as erecting some necessary 'entry gates' before rejecting the policy. Only Singapore has, to date, actually adopted a large-scale scheme of road pricing to 'sell' scarce road space during peak periods, while Hong Kong is going through a period of experimentation to assess the merits of a more elaborate, electronic monitoring and charging technology.

The Singapore system, which was introduced in 1975 as part of a wider package of measures embracing parking controls, vehicle taxation, public transport provision and infrastructure enhancement, has attracted widespread attention. The scheme is simple to operate: motorists with less than a full complement of passengers must pre-purchase an entry licence before making peak-period journeys into the city centre. Free access for (subsidised) public transport, including taxis, together with the encouragement the system offers for car pooling, keeps the cost down for non-car users while ring-roads offer circumferential routes for through traffic. Some indication of the impact on traffic in Singapore is seen in Table 3.7 — clearly congestion was immediately reduced and traffic volumes subsequently expanded at a rate significantly below that forecast. (For further details see Watson and Holland, 1977.)

Table 3.7: The Impact of Singapore's Traffic Restraint
Policies on Congestion

Time	Vehicle	May 1975 (pre restraint)	May 1976	May 1977	May 1978	May 1979
0700–0730	Cars	5,384	5,675	6,488	6,723	5,723
	Car pools	176	509	636	606	497
	Total	9,800	10,332	11,489	11,692	10,596
0730–1015	Cars	42,790	10,754	10,350	11,350	13,181
(Control	Car pools	2,369	4,641	5,337	5,684	5,756
Period)	Total	74,014	37,587	44,318	47,503	49,606
1015–1045	Cars	n.a.	6,459	6,636	6,326	5,527
	Car pools	n.a.	320	280	281	232
	Total	n.a.	13,441	13,805	14,308	15,179

Source: Seah (1980)

The Hong Kong experiment is more technically sophisticated involving vehicles being fitted with electronic identification devices which are linked to a centralised computer. When trips are made during times of high traffic congestion, the computer records the congestion costs involved and the vehicle owner is sent periodic bills in the same way as most people receive bills for their electricity or gas supply (see Fong, 1985). While technically advanced, the system has the defect that road users are not made fully cognisant of the congestion costs they generate until *after* a trip is made. The Singapore approach, while not offering the prospect of accurately reflecting the congestion costs of each individual users, has the virtue of ensuring road users know the financial implications of their trip-making activities in advance.

The political situations of Singapore and Hong Kong combined with their geographical characteristics make them convenient locations for the adoption of road-pricing types of policy. Elsewhere such convenient combinations of circumstances seldom prevail. Problems of defining appropriate zones have often been used as a practical objection to schemes *à la* Singapore while the cost of electronic metering has been cited as a drawback to the employment of more advanced technologies. Preliminary evidence from Hong Kong, however, would seem to suggest that electronic recording devices can be both practical and economic. In reality the main obstacle to the immediate adoption of road pricing is no longer technical but rather political.

There are very strong and organised pressure groups who tend to resist the notion of fiscal restraint — some because they misperceive the objectives and potential impacts but others because they would genuinely suffer a loss of welfare by its introduction (Higgins, 1981). Certainly existing road users would as a whole lose out as they have to pay to use a formerly free facility although some may gain by enjoying benefits from more rapid travel which exceed the financial outlays incurred. It is the charging agency — the central or local government — which is the immediate beneficiary of road pricing. Clearly, how the revenue collected is subsequently employed influences attitudes to road pricing and here groups such as retailers and office establishments in city centres often express concern. Retailers fear that road pricing will drive customers to other centres while employees in the controlled area fear that the charges for road use will be pushed onto them as workers seek compensatory wage rises. Generalisations are difficult here but some points may be made regarding likely future changes in attitude, both in terms of natural evolution and as the result of official policies.

The inner city areas of most Western countries have experienced decline for many years and there is an understandable fear that road pricing will accelerate this. Counter-arguments, however, suggest that in fact the opposite is more probable provided that road pricing is placed in the overall context of a comprehensive transport package involving public transport provision, road investments, parking policies etc. In most cities congestion is most acute during the morning and evening rush hours but in the near future changing patterns of work are likely to ease this pressure. The next decade or so is almost inevitably likely to witness a more intensive educational exercise on the part of transport planners as restraint policies gradually become more widespread (and, hopefully, more carefully thought out than has sometimes been the case in the past).

Many of the fears also stem from a view that the *relative* position of an urban economy would decline *vis à vis* other cities if it went alone in pursuing road pricing (a situation which exists in neither Hong Kong nor Singapore). Central co-ordination and assistance may help to allay such fears with experimental areas being compensated for any *relative* lack of vitality while, in the longer term, centrally co-ordinated policies may be required to ensure consistent adoption of restraint measures by all areas. But, come what may, it is clear that restraint will be needed in the future and the number of options left open to the authorities is now very limited indeed.

The Provision of Public Transport

The demand for urban public transport services is in decline in most Western countries despite considerable sums being provided by way of operating and capital cost subsidies. In the Third World the opposite is true with most public transport systems under increasing pressure from rapidly growing urban populations (see Table 3.1 for data on relative mode splits). The problems confronting the respective urban authorities are, therefore, entirely different. Equally, future policy approaches are likely to vary although, given the nature of the training received by many of the consultants and experts employed in Third World countries, there is a high probability there are likely to be some spillover effects from the West. It is particularly important in this area, therefore, to divide our comments clearly between those applicable to the cities of mature, developed economies and to urban areas where development is only just beginning to take off. Despite this there may still be important lessons for policy-makers to learn from the experiences of their counterparts in cities across the world.

The declining patronage, increasing financial deficits and changing operating conditions confronting public transport agencies in many Western cities have combined to produce crises. In particular, there is a growing reluctance to find the mounting sums required to pay for subsidies.

There are some exceptions and perhaps the most notable of these is Paris. France has a tradition for treating transport as a social input into the economic system and to assess its peformance in terms of its overall contribution to the economy rather than on standard commercial criteria. Public transport in Paris receives massive government subsidies and monies collected from business in the city. The system is extensive and is being expanded and modified. To date it has enjoyed a favoured position in part because of the overall French attitude towards transport (the philosophy of *droit de transport*) but also because the vitality of Paris and the dominant position of the city in the French economy made it possible to subsidise the system at least partially from local, industrial sources. Whether this will be possible in the longer term seems likely to depend as much upon the strength of the national and local economy as upon the underlying transport philosophy. Certainly, local business objected strongly in the early 1980s to additional levies to finance the system but, nevertheless, national pride in the bus and metro network is still very strong.

Apart from Paris other major Western cities seem intent in the foreseeable future on pursuing policies specifically designed to reduce the financial costs of their public transport services and to introduce greater flexibility into their provision. Rather than treat public transport as a single homogenous, all-embracing 'product' there is an increasing, albeit gradual, emphasis on tailoring services to specific markets. In part this stems from past experiences where, for example, experiments in US cities to equip all buses with elevating platforms to assist the disabled have proved both expensive and counterproductive (the time taken to 'load' wheelchair passengers and the friction generated seeming to deter the disabled from using the facility). The Singapore experiment with road pricing also revealed (see Table 3.7) that conventional public transport is not even a second-best option for those preferring automobile commuting — car pooling in particular seemed to be preferred. The trend is, therefore, towards policies which provide the specific services which either market forces suggest are in greatest demand or, where need is deemed a more reasonable criterion, meet the specific requirements of target groups.

In some areas greater variety in transport provision may sensibly be achieved by letting consumers have a greater say in what is to be provided, i.e. leave it to market mechanisms. A policy of easier market entry would have the advantage that it could well encourage more flexibility in the nature of the public transport services provided and lead to more innovative attitudes being adopted by potential suppliers. Indeed, there is ample evidence that prior to the entry regulations imposed in most Western countries a wide variety of paratransit modes operated in parallel with what we now consider conventional forms of urban public transport. In the USA, for example, 'jitneys' offered their services in many large cities up until 1915 — these usually being private automobiles which offered seats to potential travellers normally on routes into or out of central business areas. The high quality of service offered, despite high costs, made them attractive alternatives to conventional public transport (see Eckert and Hilton, 1972).

Equally, forms of paratransit currently exist in many Third World countries (for example those cited in Table 3.8) and offer important supplementary capacity to stretched mass modes of public transport. Paratransit supplies a high cost mode during peak periods (but in doing so reduces demand for both bus and tram services) and a more flexible and cheaper mode during off peaks when standard buses must run with surplus capacity. The general situation with respect to the respective costs of different modes, including several forms of paratransit, is depicted in Figure 3.6. While one may question the detail of the data, the key point which emerges is that, dependent upon the time of day and the passenger demand on a specific route, the least-cost mode is certainly not always the conventional bus. Without these supplements to bus services the transport systems of many Third World countries would come to a standstill. The difficulty in some countries is that state interest in the public transport system tends to militate against the allowance of paratransit systems. In future it seems likely that the continuation of the urbanisation process, together with increasing costs of running even a vaguely adequate urban mass transport system, will result in such Third World countries adopting somewhat less restrictive policies towards paratransit modes.

Table 3.8: Characteristics of Major Third World Jitney Systems

City	Name of service	Number of vehicles	Seats/ vehicle	Average daily ridership	Data date
Hong Kong	public light bus	4,308	8–14	1,202,000	1975
Teheran	shared taxi	12,000	5	n.a.	1970
Istanbul	dolmus	13,000	5–7	1,336,000	1970
Mexico City	pesero	17,318	6–10	1,946,000	1980
San Juan	publico	2,000	6–9	150,000	1976
Manila	jeepney	14,917	8–14	1,300,000	1970
Lagos	kai-kai	2,300	10	400,000	1974
Port of Spain	route taxi	6,000	5	120,000	1975
Beirut	service car	2,000	5	170,000	1970
Caracas	por puesto	10,035	5–15	513,000	1975

Figure 3.6: Full Cost Comparisons of Inner City Modes

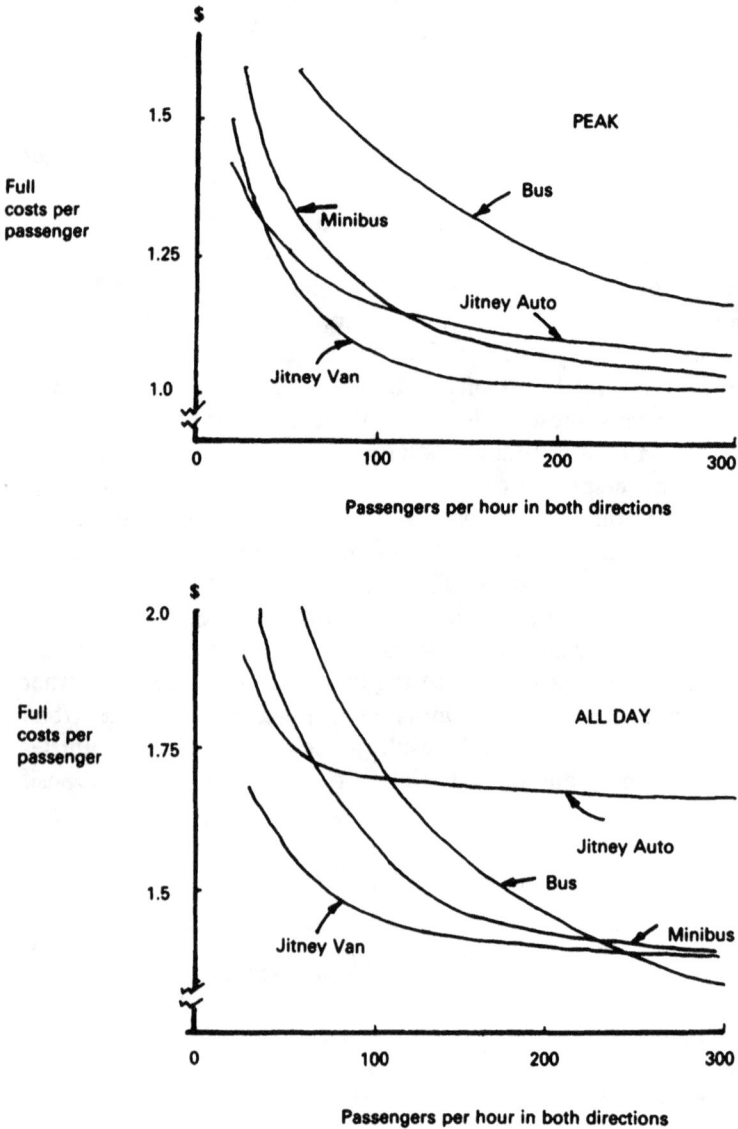

Source: derived from J. H. Boyd, N. J. Asher and E. S. Wetzler (1978), 'Non-Technical Innovation in Urban Transit; A Comparison of Some Alternatives', *Journal of Urban Economics*, vol. 5, pp. 1–20.

Market mechanisms are generally being perceived as providing a basis for much more efficiency to be injected into urban public transport. They are, however, harsh on certain groups who cannot afford the cost of such services or, for a variety of reasons, need specialised facilities. The advent of computerised information systems is, in the longer term, likely to provide a relatively low-cost solution to the problems of specific groups (e.g. the disabled, the old etc.) by being able to direct specialised vehicles more efficiently to those users requiring their use at certain times. The growth in community action groups of various kinds is currently offering important supplementary inputs into the transport available for these people and policy is gradually being modified to permit them to participate more fully and efficiently.

An emerging problem area — especially in the UK, but it also has parallels in some less developed countries — concerns the access available to those low-income households who have, as a result of the inner city development programmes which have been set in motion, removed to suburban sites. Many of these 'estates' are large, have limited recreational, retailing and employment opportunities but are located well away from the city centre. Low levels of car ownership make travel into the urban centre difficult while economic fares for bus services are prohibitively expensive given the journey distances involved. It seems inevitable that continuation of subsidy policies for public transport will be needed in such circumstances if social and political unrest is to be avoided. The key element would seem to be that the subsidies are explicit and clearly tied in with wider policies of social reform. Direct subsidies permit a closer monitoring of their effectiveness *vis-à-vis* other expenditures and also provide transport management with information to help direct their own services to routes etc. where they are most beneficial. Above all, transport once more needs to be flexible and responsive to the actual situation which is emerging.

Infrastructure Provision

The 1960s were really the years of massive investments in urban transport infrastructure. Subsequent disillusionment combined with financial stringency has, as we have seen, resulted in changes in the prevailing orthodoxy. Investment will, nevertheless, still be necessary in the future to meet both the continuing pressures for movement in established centres and the emerging pressures of the younger, expanding cities. The nature of this investment is, however, likely to differ from that of the past and the decision-making framework in which the planning of investments takes place is also likely to be of a modified form.

Much of the investment of previous periods, especially that committed to elevated track systems of the kind found in many US cities, was primarily of a functional kind, aimed at meeting the unrestrained demands of increased automobile traffic. Gradually local pressure and national legislation (e.g. on public consultation) has shifted the emphasis away from pure functionalism — aesthetics and the environment have become of at least equal importance. The onus of proof has shifted somewhat away from the community protection lobbies towards the transport lobby groupings.

The introduction of new road and surface rail systems in most older cities is unlikely to prove easy. The most obvious routings have already been adapted and while many older cities now have substantial areas of vacant or derelict land near their centres these tend to represent pockets rather than viable channels through existing buildings. A lack of confidence about both the potential success of any new scheme and the ability to insulate adjacent communities from its impact — a legacy from previous policies — is likely to generate considerable opposition to forging new links through established residential areas.

One effect of this is that increased pressure will be put on planners to invest in improving the capacity of existing infrastructure and in resurrecting systems which have gone into disuse. Some evidence of the latter approach is to be seen in the investments in the fixed-track systems in Newcastle (UK) and in Vancouver (Canada) which essentially involved the refurbishing, with some extensions, of an existing right of way. Desk-top studies have also been conducted in a number of cities into the possibility of increasing the road capacity into central areas by concreting over little-used railway lines to form bus-ways.

Many of the recent schemes to combine modern technology (usually automated light-rapid-transit (ALRT) systems) with underused or abandoned existing rights of way have represented significant advances in the way people think about mass transport provision. Furthermore they provide convenient additional capacity at relatively low social cost. Their success in meeting transport needs has been variable (see Guillot, 1984 for some views on the Canadian experiences) in part because some systems were initiated without genuine appraisal but were simply acts of faith (e.g. the ALRT of Vancouver). But also, there are clear limits as to what this type of policy can achieve. Existing rights of way are often not ideally located to meet current demands for access, especially in cities which have experienced considerable suburban sprawl.

Whether this problem of providing high quality infrastructure for radial transport purposes is likely to prove a serious long-term concern is less clear. Recent efforts to improve access to urban centres have often been founded on ideas that this would help revitalise core economies but given changes in personal aspirations and industrial practices it is becoming much less certain that this land-use planning function is realistic (see, for example, Meyer and Gomez-Ibanez, 1981). The suburban and small-city economies have grown for reasons that are unlikely to be reversed by improved transport links with older, central areas.

What is becoming clear is that policies relating to infrastructure and traffic management (including the provision of public transport services) are likely to be rather better co-ordinated than in the past. Tradition has resulted in divisions developing between the engineering-dominated investment departments of urban authorities and departments concerned with the regulation (and often provision) of transport itself. This has been a particular feature of urban administrations in the UK and North America. Pressure of scarce funding, combined with the gradual appreciation that co-ordination of strategic action may circumvent many of the pitfalls encountered in the past is already resulting in federal actions in the USA to ensure consistency in decision-rules across the two activities while centrally initiated planning requirements in the UK with respect to funding allocations are designed to improve gradually co-ordination in decision-making.

Co-ordination of a different kind is also likely to be reviewed rather more thoroughly in the future — namely that involving the roles of the various agencies that fund urban infrastructure investment. Much of the funding for major schemes tends, irrespective of the country, to come from central government. This has resulted, in many cases, in an inherent bias emerging with urban authorities having a proclivity to seeking capital-intensive solutions to many of their transport problems quite simply because the burden of payment is not *directly* from residents of their area. In the USA the distortion has been exaggerated by the extremely favourable federal financial aid available for investment in urban rail systems (including subways) and it is becoming clear that many commitments of this kind are to support services in areas where population densities are too low to yield a positive social return. (The UK system involving somewhat more balanced arrangements between central and local authorities — with the flexible Transport Supplementary Grant system — has reduced this type of problem but it still exists.) The gradual widespread acceptance of the use of consistent cost-benefit analyses of all forms of central government assistance is likely to accelerate and more sophisticated procedures are likely to be incorporated in the decision-making process as practical techniques of assessment are refined.

Policies in Third World Cities

As we saw in section 3 of this chapter, there have been two major phases of urban transport policy in less developed countries with the most recent emphasising piecemeal actions to maximise the efficiency of the existing transport systems. Increasing concern about the long-term implications of pursuing a succession of unco-ordinated short-term actions is now, however, causing some serious rethinking, particularly on the part of the main international agencies. While there seems to be some agreement that greater attention should be paid to infrastructure investments than has been the case since 1974, equally it is agreed that policies along the grandiose lines of the old land-use/transportation planning school are inappropriate. Policies part-way between the two, therefore, seem more practical and, given the influence of external policy agencies, are likely to play an increasing role in tackling the urban transport problems of Third World cities.

The exact nature of future transport policy in Third World cities is even more difficult to forecast than for developed nations. Not only are they subjected to greater political instabilities but volatile fluctuations in economic performance make it difficult to determine the potential supply of indigenous resources while the external role of both aid-giving countries and international agencies is often dominant. Comment can, therefore, only be offered with more than a modicum of caution.

A key role in future policy development must be the

appreciation that circumstances vary considerably between urban areas and, correspondingly, so should the transport policy pursued. Equally, it is now well recognised in the West that transport planning is an ongoing process which should involve long-term strategic policies and, dovetailed with these, compatible short-term plans. To achieve this necessitates considerable planning expertise, not simply periodic injections of consultancy skills. Local personnel are unlikely to be able to fulfil this role in many countries for some time to come. Policies are, therefore, likely to be developed both to integrate outside experts more fully into the planning system and at the same time to rearrange the planning administration to make better use of the local skills which are available.

The need for more adaptability amongst the 'experts' is particularly relevant to the renewed interest in providing high capacity, capital-intensive mass transit facilities. While these have clear advantages in the sense that running costs tend to be relatively low and thus reduce the problems associated with long-term funding, the land-use configurations of most cities in the Third World are not ideally suited to many forms of fixed-track infrastructure. Some imagination is, therefore, likely to be needed. Additionally, such modes of transport are likely only to act as an adjunct to much longer journeys on foot — planning must, therefore, be seen in the context of the primacy of pedestrian movements for many people.

In some cases, Third World governments have, mainly for ideological reasons, tended to limit the role played by private modes of public transit. Evidence from cities such as Bangkok (see also the discussion of jitneys in the previous section) would seem to suggest that this tends to limit important potential cost savings and that, especially in the light of severe resource constraints, the private sector might be more fully exploited with positive results. Given the need to integrate more fully the increasing number of large communities which are springing up on the edge of major urban areas but which are not currently served by public transport, flexible forms of low cost, privately run transit services may prove invaluable.

It is becoming clear from Third World countries which have been relatively successful in containing transport problems in their urban areas that one advantage they seem to enjoy is efficient administrative and enforcement machinery. It is of little use devising sensible policies if enforcement is impossible. This is particularly important in cases where rising levels of automobile ownership are creating severe congestion conditions or where transport is impeded by non-transport activities. The policy-makers can attempt to improve enforcement and encourage compliance by spending more on policing and programmes of education but in reality the effects are unlikely to be more than marginal. As the overall development process continues, however, and social systems of control and co-operation become more regulated, so to a large extent the problems of compliance will recede — this is an evolutionary process which policy-makers and planners can only assist in.

References.

Anderson, S. C. (1983), 'The Effect of Government Ownership and Subsidy on Performance: Evidence from the Bus Transit Industry', *Transportation Research* (Series A), vol. 17, pp. 191–200.

Cooter, R. and G. Topakiam (1980), 'Political Economy of a Public Corporation: Pricing Objectives of BART', *Journal of Public Economics*, vol. 13, pp. 299–318.

Department of Transport (1984), *Buses*, Cmnd 9300 (HMSO, London).

Downs, A. (1978), 'The Law of Peak-Hour Expressway Congestion', *Traffic Quarterly*, vol. 9, pp. 622–32.

Eckert, R. D. and G. W. Hilton (1972), 'The Jitneys', *Journal of Law and Economics*, vol. 15, pp. 292–325.

Elliott, J. R. and C. C. Wright (1982), 'The Collapse of Parking Enforcement in Large Towns: Some Causes and Solutions', *Traffic Engineering and Control*, vol. 23, pp. 304–10.

Fielding, G. J. (1983a), 'Changing Objectives for American Transit. Part 1. 1950–1980', *Transport Reviews*, vol. 3, pp. 287–99.

Fielding, G. J. (1983b), 'Changing Objectives for American Transit. Part 2. Management's Response to Hard Times', *Transport Reviews*, vol. 3, pp. 341–62.

Fong, P. K. W. (1985), 'Issues of the Electronic Road Pricing System in Hong Kong', *Transportation Planning and Technology*, vol. 10, pp. 29–41.

Fothergill, S. and G. Gudgin (1982), *Unequal Growth: Urban and Regional Employment Change in the UK* (Heinemann, London).

Guillot, E. (1984), 'Bus Transit Interface with Light Rail Transit in Western Canada', *Transportation Research*, vol. 18, pp. 231–41.

Higgins, T. (1981), 'Road Pricing: A Clash of Analysis and Politics', *Policy Analysis*, vol. 7, pp. 71–89.

Maunder, D. A. C. (1983), *Public Transport in Relation to the Travel Needs of the Urban Poor in Cities of Developing Countries*, Ph.D Thesis, University of Leicester.

Meyer, J. R., J. F. Kain and M. Wohl (1965), *The Urban Transportation Problem* (Harvard University Press, Cambridge, Mass.).

Meyer, J. R. and J. A. Gomez-Ibanez (1981), *Autos, Transit and Cities* (Harvard University Press, Cambridge, Mass.).

Ministry of Transport (1963), *Traffic in Towns* (HMSO, London).

Monopolies and Mergers Commission (1982), *Bristol Omnibus Company Ltd., Cheltenham District Traction Company, City of Cardiff District Council, Trent Motor Traction Company Ltd. and West Midlands Passenger Transport Executive — A Report on Stage Carriage Services Supplied by the Undertakings*, HC 422 (HMSO, London).

Morrison, S. A. (1986), 'A Survey of Road Pricing', *Transportation Research*, vol. 20A, forthcoming.

O'Neill, T. J. (1977), *On Illegal Parking in Boston* (Office of the Collector-Treasurer, Boston).

Ortner, J. and M. Wachs (1979), 'The Cost-Revenue Squeeze in American Public Transit', *Journal of the American Planning Association*, vol. 45, pp. 10–21.

Reid, R. B. (1983), 'Subsidies in British Transport', *Transport*, September/October, pp. 20–1.

Reynolds, D. J. (1960), *The Assessment of Priorities for Road Improvement*, Road Research Laboratory, Road Research Technical Paper 14, Crowthorne.

Schaeffer, K. H. and E. Sclar (1975), *Access for All: Transportation and Urban Growth* (Penguin, Harmondsworth, Middx.).

Seah, C. M. (1980), 'Mass Mobility and Accessibility: Transport Planning and Traffic Management in Singapore', *Transport Policy and Decision Making*, vol. 1, pp. 55–71.

Sharp, C. H. and A. Jennings (1976), *Transport and the Environment* (Leicester University Press, Leicester).

Thomson, J. M. (1974), *Modern Transport Economics* (Penguin, Harmondsworth, Middx.).

Thomson, J. M. (1977), *Great Cities and their Traffic* (Victor Gollancz, London).

Thomson, J. M. (1983), 'Towards Better Urban Transport Planning in Developing Countries', *World Bank Staff Working Paper*, no. 600.

Wachs, M. (1981), 'Pricing Urban Transportation: A Critique of Current Policy', *Journal of the American Planning Association*, vol. 47, pp. 243–51.

Watson, P. L. and E. Holland (1977), *Road Pricing in Singapore: Impacts of the Area Licence Scheme* (World Bank, Washington).

Wingo, L., Jr. (1961), *Transportation and Urban Land Use* (Resources for the Future, Washington).

Young, K. and E. Mills (1983), *Managing the Post-Industrial City* (Heinemann, London).

4 THE REGULATION AND CONTROL OF TRANSPORT

Regulation and Public Policy for Transport

Regulation and control of transport is currently one of the most contentious areas of public policy. In both Britain and the United States it has become something of an obsession, figuring prominently in the current work of the European Council of Ministers of Transport (ECMT, 1985). There is little likelihood that the issues will die away.

One reason for this current concern has to do with state-owned or -managed enterprises which have achieved, it is contended, a direct or quasi monopoly status. It is perhaps unsurprising, therefore, to find that transport operations figure prominently, given the ways in which transport policy has evolved and the structural changes that have been wrought in the sector. With the shift in emphasis following the demise of the Age of Administrative Planning (see Chapter 2), together with a redefinition of the public interest and a reassertion of the merits of allocative efficiency, the role of state-based regulatory regimes has been increasingly questioned.

This assault on regulation has brought into sharp relief two contrasting approaches to transport policy, one arguing that state intervention is a prerequisite to any effective policy, the other that intervention is the obstacle to effective policy. The arguments for intervention have already been noted in Chapter 2 but here they are taken a stage further, by examining the rationales behind regulation and deregulation, using case-study material mainly on buses but also railways and aviation drawn from contemporary North American, British and European experiences.

Recent attention in public policy has focused on monopolistic competition and the role of state regulation in either supporting or reacting against monopolistic practices. Most countries in the Western world have the equivalent of the British Monopolies and Mergers Commission or the US Federal Trade Commission and Antitrust Division of the Justice Department. In these agencies the main concern is with the corporate behaviour of trading companies. The guiding principle always has been to protect the public interest in matters where monopolistic practices are evident or potentially in evidence. Such agencies have legal powers to stop acquisitions or trading practices which are deemed to be against the public interest.

It is against this background that arguments have raged, particularly on two quite different aspects of monopolistic behaviour which often get confused. The first concerns the problem of what to do about monopoly structures, the second with the issue of monopoly power. Monopoly structure refers to those types of production which are vested, more often than not, in very large single-product organisations where indivisibilities are present and large-scale production is by a single supplier. Examples found in textbooks often refer to electricity supply, where one generating plant can produce sufficient electricity to meet very large demands. This natural monopoly of supply is seen as the most efficient way of meeting consumers' demands subject to certain caveats, which is where public interest arguments have traditionally been used to justify some form of regulation. So as to ensure that such a natural monopoly does not abuse its position by exploiting consumers, the convention has been for the state to appoint a public agency to oversee pricing policy. In return for a legally binding agreement to have their pricing policy and practices subjected to regulation, the regulators have traditionally protected established suppliers from new entrants to the market.

This approach has characterised the regulation of public transport in North America. The British experience has in a way been similar, but instead of a protectionist approach to the public interest, the analysis of the 1930 Road Traffic Act showed that the creation of a quasi-monopoly in public transport was a clear policy objective — in order to limit wasteful competition, to reap the economies of scale anticipated from large-scale mergers, and to enhance public safety. British transport policy therefore actively fostered monopolistic supply, albeit subject to a complex regulatory regime. In both North America and Britain, existing and established suppliers were granted grandfather rights in exchange for protection from new entrants.

The argument thus far has examined monopoly structures; however, it is also possible to begin an analysis with monopoly power, which may or may not lead to monopoly structure. Monopoly power implies that a single producer or supplier has the ability to make a price which it knows will stick. Monopoly structure does not guarantee this. Public passenger transport traditionally exhibited the features of a monopolistic structure but in reality it has been confronted by competition from the private car. The relationship between public transport and motor manufacturers in the USA is an enlightening one, illustrating the use of monopoly power.

Of the 20 largest industrial corporations in the United States, General Motors (GM) ranks first. Of that 20, 12 have a major interest in automobile and related production, including names like Exxon, Ford, Chrysler, Mobil Oil, Texaco, Gulf Oil, US Steel, and Goodyear Tire and Rubber. This group accounts for 71 per cent of all sales, 72 per cent of total assets, 68 per cent of net income, and 57 per cent of all employees. General Motors' share of these 14 automobile and related corporations is striking and illustrates not monopoly structure but monopoly power: GM accounts for almost a quarter of total sales in the industry, nearly 16 per cent of the industry's total assets, 27 per cent of total net income, and almost 34 per cent of employment. As Taebel and

Table 4.1: Ownership of US Automobile Facilities at Each Level of Production

Components production	No. of plant
GM	73
Ford	39
Chrysler	31
American Motors	5
Total	148

Final assembly	
GM	22
Ford	15
Chrysler	6
American Motors	1
Total	44

Retail distribution (US makes)	
GM	12,045
Ford	6,684
Chrysler	5,406
American Motors	1,947
Total	25,427

Source: Taebel and Cornehls, 1977.

Cornehls (1977) put it, 'the great corporations that have emerged to fill the transportation needs of American Society have put together the most imposing concentration of private economic power the world has ever seen'. Table 4.1 illustrates this by the three major activities of ownership: components production plant, final assembly plant, and retail outlets.

The history of General Motors' commercial practices is not particularly savoury, even more so when its involvement in public transport is examined. With GM's rise to economic prominence came the ability to make prices rather than to take them from the market place. Part of GM's evolving corporate strategy was to devise means to eliminate viable alternatives to the private car. GM's history is studded with instances where their system of administered prices, and the super profits which went with it, allowed them to buy out competitors or force them out of the market. The attempt to eliminate competition took place in three phases: the substitution of bus for rail passenger transport; the acquisition and dismantling of local electric systems, to be replaced by buses; and the eventual displacement of GM-controlled bus systems by private cars produced predominantly by GM. For example, in 1925 GM acquired the country's largest bus producer; in 1926 it participated in creating the Greyhound Corporation (whereby Greyhound agreed to purchase almost all its buses from GM, who agreed not to sell any buses to Greyhound's competitors), and remained its largest shareholder until 1948. In times of financial hardship, GM lent Greyhound monies to ease cash flow problems.

A similar process characterised GM's involvement in the purchase, ownership and control of local electric systems and bus manufacturers. United Cities Motor Transit, a GM-formed holding company, was created solely to purchase electric streetcar companies, convert them to GM-built bus systems, and resell them to local operators subject to agreements granting GM the monopoly supply of equipment. Beginning in 1936, New York City's streetcar system was converted to GM buses in just 18 months. By 1949, GM had participated in the elimination of over a hundred electric systems in 45 US cities. As Taebel and Cornehls (1977) says, 'by the close of the 1940's, GM's devastating assault on the electric railways was virtually complete'.

GM also had effective monopoly in bus manufacture, as well as a major interest in railways (see Table 4.2). By a process of mergers and rationalisations, only one independent producer was engaged in bus manufacture by 1973 — and that company was controlled by a Vice President of General Motors. The table also makes a comparison between US producers and manufacturers of similar equipment in three other countries with similar population levels. In the United States there were less than 10 per cent of the number of companies operating in France, West Germany and Japan — markets with similar characteristics to the USA.

Table 4.2: Bus and Rail Industries. Total number of producers in United States and West Germany–France–Japan

	United States	*West Germany–France–Japan*
Buses	3	26
Railcars	3	23
Train locomotives	2	33
Total	8	82

Source: Taebel and Cornehls, 1977.

Much of the current debate on regulation is North American and is based on critiques of established regulatory regimes which have attempted to 'do something' about monopolistic practices. Traditionally the policy response has been either to do nothing about private monopolies, relying on the legal redress of abuse through the courts when it becomes apparent, or to create a regulatory regime to oversee and control such ventures. In North America the tradition has been to pursue both policies, although public transport has, almost without exception, been subject to the regulation approach. In Britain the response has been similar, but taken one step further — public ownership of those ventures which are deemed to be significant for national policy purposes. In every case public-interest-type arguments have been employed which, in the case of monopoly power, try to influence the market conduct of suppliers and hence their performance. The objectives have been threefold: to restrict the potential ability of suppliers to engage in predatory pricing strategies, because they may lower prices beneath costs to drive out potential competition; to reduce the prospect of collusion between potential rival suppliers, in order to force out competitors; and to disallow direct trading agreements between potential suppliers, in order to capture the market. In other words, each objective is concerned with the problem of market failure.

Market conduct and market failure are just two of the reasons traditionally advanced for introducing and maintaining regulatory regimes in transport. But regulation has been linked historically, as we have seen, to other policy objectives, to do with combating wasteful competition and ensuring a high level of safety. The policy instrument selected was almost without exception one of assisting the process of the development of monopolistic structures for the supply of transport in all modes. These structures were deemed to be in the public interest; regulation only came about to ensure the protection of the public interest, should the potential power inherent in monopoly supply be abused.

The two traditional objectives which have underpinned the regulatory theme in transport policy merit attention, because it is around these that the current moves for deregulation have been based. Put simply, regulation has attempted to control both the quantity of transport supplied and its quality, although the policy objectives have tended to treat both simultaneously. The first objective concerns the need to avoid wasteful competition. Many economists see a competitively organised market of many small suppliers within a particular transport mode as both a desirable and necessary condition for an efficient level of service provision; the same logic is applied to markets where there are competing modes, for example, between long distance rail and aviation. The proponents of regulation do not share such a belief in the untrammelled virtues of competition but only because they see it as just one part of a more complex equation which has to allow for other factors to be included. The point has been discussed at length by Peter Self (1975) who distinguishes between the 'economic rationality' of market economists and what he calls the 'political rationality' of the political process, and the state of creative tension which exists between them.

The political rationality view of wasteful competition, as stated for example in the debates surrounding the passage of the British 1930 Road Traffic Act and the subsequent performance of the Traffic Commissioners, is that unregulated markets for transport are simply not in the public interest, because they do not allow for stability in the level of service or fares/rates, nor do they permit an adequate degree of co-ordination between services and modes. These are terms which are undoubtedly ambiguous and debatable, and are not ones which are central in the orthodoxy of economic rationality. For example, Glaister and Mulley (1983), in their study of public control of the British bus industry, state that there is only one 'rationalisation of the commonly used phrase "wasteful competition" that we know of', and that concerns a multi-product natural monopoly supplier which may become 'unsustainable' due to the threat of competition. But notions of stability and co-ordination are fundamental axioms of political rationality. Political rationality treats economics and economists as important parts of its armoury, but assigns them a particular role — to analyse the potential economic consequences of a particular set of policy choices, and to justify those choices in economic terms.

In the case of transport policy, regulation is involved to create a semblance of stability and co-ordination in what are perceived to be non self-regulatory markets. The traditional public interest solution sees virtue in creating and sustaining a regulated market through fostering a monopoly structure, with external regulatory control exercised by an administrative body. The traditional economic justification is that by allowing the development of monopoly structure, the potential benefits accruing from the increasing returns to scale thus generated are in the public interest. The assumption which is asserted is that by encouraging mergers and rationalisations subject to regulatory control, stability in and co-ordination of public transport can be achieved.

The second objective underpinning traditional arguments about regulation concerns the quality of supply and the quantity of service necessary to ensure quality. In practice two issues have dominated transport policy — public safety and information. Public safety arguments concern the need to protect individuals travelling on public transport and to protect people from the danger posed by public transport. Historically this approach to regulation and control cannot be underestimated. In practice such controls cover the use and operation of public vehicles through a complex licensing system which includes regulations on the design and manufacture of vehicles, regular vehicle inspections, tests of drivers' competence, and the setting of minimum design standards for infrastructure. Such aspects of market failure are dealt with summarily by those supporters of economic rationality: safety regulation is seen as a necessary requirement, imposing additional costs on operators which they would otherwise not recognise. In exchange for this, regulators offer a degree of monopoly rights by way of compensation. The political rationality view of safety regulation is different. In order to achieve a minimum acceptable level of safety the behaviour of transport operators must be subject to basic legal and administrative rules governing the use and construction of vehicles. This can only be achieved by dealing with a relatively small number of operators who have already demonstrated that they take their responsibilities seriously. To achieve required improvements in safety it is necessary therefore to control the number and quality of operators, and hence their entry.

For competitive markets to operate in their pure form, a further crucial assumption has to be met — that perfect information is freely available. This is not to deny the crucial role which risk and uncertainty plays in clearing markets, but to emphasise the point that they are a logical consequence of competitive behaviour. The key point from the transport policy perspective concerns the sort of strategies which can be devised to manage risk and uncertainty. Once again we have two very contrasting views. From the economic rationality viewpoint, risk and uncertainty are integral elements of a competitive public transport service, and consumers should not come to expect stability with respect to reliability, fare levels and co-ordination between competing operators. We must learn to trust the market and understand that in the long run a strategy based on the exploitation of risk and uncertainty will yield overall higher net social benefits. If there is operator failure, for example in reliability of service, then consumers will soon signal their unhappiness and the market will respond by the entry of new operators. Market failure, in terms of a fundamental inability to provide the required level of information, is more often than not left unconsidered (cf. Hibbs, 1982; Department of Transport, 1984).

It is distrust with this risk exploiting strategy which lies at the heart of the political rationality approach to information. In terms of transport policy, the response is couched in a way which implies that public transport is too important to be left to the market. If there is an economic rationale behind this view, then it is that the transaction costs associated with a market solution to public transport provision are prohibitive. The assertion is that the levels of information availability guaranteed, as a result of regulation, a level of service which a multiplicity of small competing operators could never achieve. Information is seen as a central means of guaranteeing stability both of operation and in level of service. Risk and uncertainty are not denied, but the approach to strategy is to avert or at least minimise their impact.

The objectives behind regulation and arguments for deregulation are heavily debated. The notion that competition in public transport is both wasteful and undesirable is a commonly held view which cannot be put down solely to special pleading by monopolistic interest groups. Similarly there is a widely shared view that public safety and information require a special kind of regulation to protect the public interest. The real area of controversy is between the two dominant rationalities which underpin transport policy. To complicate matters, however, there is one further consideration that cannot be ignored — the twin issues of industry structure and public subsidies.

As noted previously, governments intervene in transport provision to secure a variety of different and competing objectives. In the early part of this century, one of the more pressing concerns was the decline of rail transport. This became reflected in arguments about the evolving structure of the transport industry, and the fears that the growth in small independent transport operations would seriously erode attempts to revitalise the railways and create unacceptable problems for public safety. At that time, and in all countries where this process was accelerating, the policy response was very similar: to attempt to restructure the emerging industry away from its path of a multiplicity of competing small independents towards a smaller number of enterprises which would approximate to localised natural monopolies. This strategy was adopted for a clearly defined purpose: the fostering of localised natural monopolies which would, on the one hand, reap the expected benefits of increasing returns to scale and, on the other, enable the restructured transport industries to be regulated in the the public interest. It is interesting to note that these were also the arguments used when the British government attempted to nationalise public transport in the 1940s. Equally interesting is that this solution is one of the very few examples where the competing arguments of the two rationalities appear to have coalesced.

The critical assumptions made at that time were at best guestimates of how the transport sector would shape up in a regulated environment. The belief was that by inducing (or not restricting) concentration in the industries, and controlling entry into the respective markets, a process of increasing returns to scale would be set in motion.

In the intervening years, arguments have raged as to whether or not public transport industries exhibit increasing returns to scale, and despite a vast literature on the subject there remains little unanimity (e.g. see Nash, 1982). If the evidence is inconclusive, what is known is that, in terms of political rationality, the regulatory regimes have until recently retained their policy appeal in the developed world. Given the central requirement for generating stability, these regimes have embraced an administrative philosophy which includes uniformity (of fares structures, levels of service, etc.) across geographical areas and through time. The question being asked with increasing frequency is, if such services are to be retained and standardised fare structures continued to be applied, whose responsibility is it to meet the rising financial shortfalls which are confronting the operators?

In practice, subsidy is met from two main sources: from operators' surpluses, in the form of cross-subsidy of unprofitable services by profits made elsewhere, and from revenue support by central and local government. The degree of internal cross-subsidisation in the public transport industries is a notoriously difficult subject. Lack of information, however, has not precluded those supporters of allocative efficiency from roundly condemning it as a practice which should not be tolerated; as a recent British policy statement declares: 'Competition and tighter financial constraints on incumbent operators will result in a much lower level of cross-subsidy . . . On the basis of the limited evidence available it cannot be claimed that such changes are generally undesirable' (Department of Transport, 1984).

Subsidies to operators from public sources, however, are much more visible. What the evidence shows is particularly stark, since in virtually all countries since the 1970s the level of public support has been increasing consistently (see also Figure 3.2 in Chapter 3). In the UK, the proportion of subsidy to total operating cost has been estimated at about 8 per cent in 1970; this increased to around 27 per cent by 1980. In the United States, comparable estimates are 15 per cent in 1970 increasing to 60 per cent in 1980; in the Netherlands, 48 per cent rising to 86 per cent; and in West Germany, 23 per cent to 40 per cent. In a crude league table of levels of subsidy, in 1970, the UK and USA were 8th and 7th respectively of 13 countries; by 1980 the UK was 11th and the USA was 5th (Webster, 1985).

In Great Britain revenue suport for buses alone has increased from £31 million in 1972 to £435 million in 1982 (at constant prices), or 0.4 pence per passenger journey in 1972 increasing to 7.9 pence ten years later. By 1983, subsidies accounted for 38 per cent of operators' income. In one local area, South Yorkshire, where the promotion of public transport has been a major policy objective of local government, subsidies accounted for an estimated 80 per cent of income in 1982, or 13.7 pence per passenger carried.

The reasons behind these figures require some examination, if only because one of the main arguments of the proponents of deregulation is aimed at what they see as unacceptably high levels of subsidy. Their most significant concerns are the continuing decline in public transport patronage and increasing operating costs.

In Great Britain, for example, in the ten year period from 1972, the number of bus passenger journeys fell by over 30 per cent and total receipts by more than 12 per cent. In the same period fares increased by about 30 per cent, the number of households with regular use of a car increased by almost 10 per cent, and those with two or more cars almost doubled. In 1973, 77 per cent of passenger transport miles were by private motor; by 1983 this had increased to 83 per cent. In the ten year period, passenger miles accounted for by the bus fell from 13 per cent to 8 per cent and rail transport fell from 9 per cent to 7 per cent. Taken together the annual rate of patronage decline has been estimated at about 4 per cent.

On the other hand, the picture in the United States seems rather different. In the same period, 1972 to 1982, intercity bus travel actually increased by 5 per cent and the share of private motor travel fell from 87.1 per cent to 83.5 per cent of total passenger miles. The share of bus and rail travel remained remarkably constant with air travel increasing from 9.5 per cent to 13 per cent. Overall, and largely because of the increase in air travel, the share of public passenger transport increased from 12.1 per cent to 15.6 per cent of total passenger miles.

A recent study has examined the impact of subsidies in 17 countries, exploring the effect of eliminating subsidy altogether and the amount of subsidy required to stabilise patronage at current levels (see Webster, 1985). Under existing regulatory regimes and subsidy arrangements, trends in patronage ranged from an average annual increase of 3.72 per cent in Italy and 3.58 per cent in Canada to a decrease of 4.08 per cent in Great Britain and 2.95 per cent in New Zealand. It is estimated that had the rate of subsidy not increased, these trends would on average have been a little less than 2 per cent lower per year.

One can take the analysis a stage further by attempting to estimate the impact of the complete withdrawal of subsidy. Taking all 17 countries it is calculated that patronage would have fallen by 29 per cent over a 13 year period and that fares would have risen by 96 per cent. In Great Britain the complete withdrawal of subsidy in 1965 would have meant a drop of 23 per cent in passengers and an increase of 46 per cent in fares, while in the United States over the same period patronage would have fallen by 42 per cent and fares increased by 164 per cent. The worst picture is reserved for the Netherlands: patronage would have declined by 50 per cent and fares would have increased by 274 per cent.

To complete this analysis, one can look at the likely trend in bus use and subsidy for Great Britain to the turn of the century. Assuming that policy was to maintain patronage at its 1982 level, subsidy would have to be doubled in real terms. If, however, the policy objective was one of maintaining subsidy at its 1982 level in real terms, then patronage would fall by about 20 per cent.

The problem this analysis highlights from a transport policy perspective, however, is not whether the calculations are correct, but rather the dilemma facing those responsible for transport policy, based on the contrasting advice which flows from an appraisal of its consequences. For example, if one accepts the arguments of the proponents of economic rationality, then the policy objective is predicated on the requirement to reduce long-term, open-ended subsidy commitments. The consequence which follows from the above analysis is a further reduction in patronage, and a switch at the margin from public transport use to personal car ownership. The principal role under these circumstances is accorded to the market mechanism to solve the mix of journeys and modes demanded, and regulatory regimes are seen as obstacles to this efficient process.

Contrast this line of reasoning with those of the political rationality approach. Here the policy objective may have little to do with allocative efficiency type arguments, since the political objectives of regulatory regimes and subsidy arrangements may be proxies for more far-reaching aims: 'keeping fares low' or 'promoting public transport'. Ambiguity and imprecision are significant political considerations. A long-term, open-ended subsidy commitment in this sense may not be economically efficient but it can be politically efficient, if only because it demonstrates a commitment to those households without regular use of private transport that their transport requirements are on the political agenda (see Chapter 7 for a more detailed analysis).

Regulatory regimes may also assume a crucial importance in securing political authority and no more so than in those spheres of civil transport which impinge on defence and military operations. Perhaps the best example of this is provided by Paris in the last century, and particularly the wide Boulevards which have become such a hallmark of the city's attractiveness. The investment in the Boulevards, along with the replanning of Paris itself by Haussman, was undertaken primarily for military purposes — to ensure the rapid deployment of troops throughout the city to quell civil unrest.

What these arguments suggest, is that neither economic nor political rationalities have an automatic right to dominance in matters involving transport policy. Neither is right nor correct, nor does it make much sense to envisage a situation where one could replace the other. What exists is a situation of tension between the two, characterised by periods where one or the other is more dominant. In the Age of Administrative Planning it is clear that variants of political rationality were in the ascendancy. Today the picture appears to have been reversed. However, it would be a mistake to assume that there is a cause–effect relationship at work, because the two rationalities are mutually dependent. Arguments about the merits and demerits of regulation have always been present, and in all probability will remain into the foreseeable future but not necessarily in the form in which we see them today.

Regulation and Deregulation: Some Modal Analyses

Buses

The late 1970s saw the first real moves on the part of those responsible for regulating the bus industry to review their respective regimes. In Britain, this was the first concentrated effort to rethink the value of and role for regulation since the passing of the

1930 Road Traffic Act. In the United States, a formal review was begun, the first significant re-evaluation since the passage of the 1935 Motor Carrier Act. In both countries the first moves to alter fundamentally the legislative arrangements for regulating the bus industry were directed at interurban routes: in Britian these were enshrined in the 1980 Transport Act, and in the US the 1982 Bus Regulatory Reform Act. The objectives behind both Acts were remarkably similar: to introduce greater competition into the industry by liberalising the regulatory regimes, particulary with regard to price setting and quantity controls. Whether this amounts to deregulation in the real meaning of that word is questionable, since in both countries the administrative bodies previously responsible for regulation remain. Equally, in the US, bus operators have retained immunity from prosecution under the antitrust laws which allows them to continue collective pricing behaviour through the interstate bus rate agency. In Britain, the 1980 Act made substantial changes to licensing arrangements, but retained the overall regulatory environment of public service vehicle operator's licences and road service licences.

So, at first glance, it seems that initial moves to deregulate transport involved less than radical changes. A more considered appraisal, however, suggests a different conclusion. In both pieces of legislation the most significant change concerned the protection of existing established operators with the 1930 and 1935 Acts both granting grandfather rights to established operators. In Britain established operators had the automatic right to challenge any application for a new route licence, with the onus resting on the applicant to prove the merits of the proposal. In the USA a similar process applied: applicants for a new licence — an interstate motor common carrier authority — had to demonstrate that the operations they proposed 'are or will be required by the present or future public convenience and necessity'. As early as 1936 the Interstate Commerce Commission (ICC) defined this to mean 'whether this purpose can and will be served as well by existing lines or carriers; and whether it can be served by the applicant . . . without endangering or impairing the operations of existing carriers contrary to the public interest'.

In both countries it was this onus of proof which was changed in the 1980 and 1982 Acts — the onus of proof is now with the regulatory agency and the established operators to demonstrate that it is in the public interest *not* to grant an application. Both

Acts therefore reduce the weight previously accorded to the views of established operators.

It should be borne in mind that this legislation was directed only at the express services or interstate operators. It was not specifically aimed at urban or rural-to-urban services, with one important exception. In Britain the 1980 Act introduced Trial Area designations, in which any local authority could apply to the Minister to have a geographical area designated as a Trial Area where road service licences would not be required. This was an attempt to experiment with the abolition of quantity controls on the number and frequency of buses running on a local network. So long as an operator could meet the requirements for a public service vehicle licence, he or she could operate on any route subject only to the local authority being notified in advance about the proposed new service, and the service being advertised in advance. These experiments were intended to 'demonstrate whether the interests of the public would be served by a complete abolition of road service licensing'.

Clearly the transport policy-makers must consider these experiments to have been successful, because five years later the government has passed a new Transport Act which abolishes road service licensing for conventional bus operations throughout the country. In the *Buses* White Paper (Department of Transport, 1984), the minds of the administration seem to have been won by the evidence: 'So the abolition of road service licensing can provide real benefits in rural areas.' In the same document, however, in an analysis of the effects of the Trial Area schemes, the conclusions reached by those responsible for reviewing the consequences are much less sanguine: 'Deregulation has not been sufficient to halt the overall decline in rural bus services, but neither has it worsened the situation. Operators have benefited marginally from simplified administrative procedures.'

But what of the evidence on which these two rather contradictory views were based? The Hereford Trial Area (see Evans and Hoyes, 1984) was important not only because it was the only scheme which included a sizeable urban area, but also because the local authority with responsibilities for public transport subsidies applied a new approach to revenue support, based on a system of competitive tendering for unremunerative routes, and because there was a substantial presence of independent operators already in the area. The local authority tendering system operated mainly in the rural areas, and the limited competition took place in and around the urban area of Hereford itself. The evidence of tendering for subsidy on unremunerative routes is striking: of the 53 services which no operator seemed prepared to run without subsidy, 15 were run by a subsidiary of the nationalised NBC which decided to continue 12 of them without subsidy. There was direct competition by tender for 32 of the services, resulting in 19 changing operator. The local authority estimated that as a result of this tendering, an annual saving of £62,000 was made in revenue support, coupled with additional savings of £65,000 in other public transport operations where it had statutory obligations, e.g. provision of school buses (see Department of Transport, 1984). Because of its policy of not applying for subsidies by tender, the NBC subsidiary's level of actual subsidy paid fell to zero. One consequence of this was that the proportion of subsidy-free bus-miles increased from 22 to 69 per cent; the equivalent independent operator's proportion increased from 43 to 61 per cent. In total only 20 per cent of bus-miles were subsidised following the Trial Area contracts. Those services which changed operator accounted for less than 4 per cent of bus-miles. The contracts were awarded initially for 18 months and the second round of tendering in 1983 resulted in 38 new contracts, the evidence suggesting that competition had strengthened during the period. Following the second round, the proportion of subsidised services decreased further.

The evidence of changes in bus provision in the urban area indicates a high degree of instability. The changes in the subsidy arrangements have had little impact, since the vast majority of services changed to a subsidy-free operation. Because of the decision of a major NBC subsidiary not to participate in tendering, the immediate impact had been a reduction in services of about 20 per cent. One new independent operator entered the market with the aim of replacing those lost services but this was not sufficient to bring the level of service back to its previous position. Three other independents began providing urban services within the first 18 months, mainly in direct competition with the established operators. There was little innovation, though, either in the provision of new routes or in the style of operation.

Bus patronage, however, increased significantly, mainly because of the improvement in the frequency of services along established routes. One estimate puts this at 66 per cent. Competition seems to have had most immediate effect on fare levels. The typical fare level before the start of the Trial Area was about £0.32p; within two years this had fallen to about £0.10p (Fairhead and Balcombe, 1984). In a follow-up study, Evans and Hoyes have estimated that bus-miles increased by over 50 per cent and of this the independents' share is estimated at over 70 per cent.

The combination of new entrants, lower fares and higher service provision offers at first sight tangible evidence to support the government's policy of local competiton. But if these are the benefits to the consumer of bus travel in Hereford, then we must also consider the costs of achieving them. As official monitoring shows, there has been an increase in bus patronage: 'but not sufficiently to support all the additional buses, especially at the current low level of fares. Surveys . . . indicated average (weekday/daytime) revenues of half the . . . average cost per vehicle mile' (Department of Transport, 1984).

In other words, all operators were running services which failed to meet operating costs. Using a naive economic model of bus operations, it has been estimated that immediately prior to Trial Area designation the NBC operations were breaking even and the independents were making a small operating surplus. Within two years of designation, the public sector operations were in deficit by a little short of £500,000, and the independents' combined deficits were well over £200,000. Although evidence of this type must be treated with care, it does indicate the financial cost of operating a level of service of the kind which produces the user benefits noted earlier.

An indication of operators' responses to this situation can be gauged from some of the events which have occurred in the Hereford Trial Area. By 1985, in terms of urban services, fares have been increased on average by little short of 100 per cent in two years, service frequencies have been reduced significantly, and the NBC subsidiary has gained a level of local monopoly which it had not enjoyed even in the pre-Trial Area years. The major independent operator prior to the experiment sold out to the NBC, and of the other four independents who entered the market, not one remains with a significant presence.

A similar story can be told for the deregulation of the UK long distance coach market. The official policy is that 'the legislation has certainly been justified: the results have been dramatic'. From the coach user's viewpoint, there have clearly been benefits, in terms of lower fare levels and an increase in the level of service:

mainly higher quality luxury coaches, the introduction of almost 700 new services, and a veritable explosion in the marketing of services. Set against this, however, the size of the market has remained reasonably stable at between 17 and 18 million passengers per year, even though fare levels have been reduced significantly. The result has been a fall in total revenue whilst operating costs have increased. The major competitors in the new deregulated environment were a subsidiary of NBC (National Express) and British Coachways, a consortium of six established independent operators, mainly because they operated a national network of services. (Although many small operators did enter the market, these tended to be on a route basis rather than through the provision of a network.) Competition between the two national network operators was intense, and proved to be a contributory factor to the eventual disbanding of British Coachways within a year of operation. One of the reasons for the high failure rate of both small independents and the national consortium can be attributed directly to two of the major strengths of the established public operators — a network of travel agents handling bookings, and the presence of bus stations.

After an initial burst of new entrants, accompanied by an equivalent rate of exit, the public sector operation has reasserted its dominance in the market. One estimate suggests that immediately prior to deregulation, public sector operators accounted for about 65 per cent of passenger journeys and independents 35 per cent. These shares were maintained in the year following the 1980 Act. However, by 1982 the public operators had increased their market share by 12 per cent to 72 per cent and by 1983 to 76 per cent, largely attributable to a 6 per cent fall in passenger journeys made using independent operators.

Four years after deregulation a new pattern had emerged which emphasised several distinct styles of express operation in addition to the network of National Express. Firstly, there had been a marked increase in the number of independent operators offering a higher level of luxury long-distance coach travel; secondly, a number of independent operators ventured into collaborative agreements with National Express to run joint services; and thirdly there had been the development of long-distance commuter coaching, especially into London.

There are also wider consequences of the regulatory reforms. Although difficult to quantify, it is generally agreed that as a result of the 1980 Act, and the pressures for operational efficiencies thus generated, some previously served locations on express routes have been severed — particularly stopping points in smaller settlements. Also some less profitable cross country routes have disappeared. A second point concerns the impact of intra-sector competition with other modes. In the case of long-distance travel, it is clear that British Rail's revenues have been the hardest hit, particularly its Inter-City traffic. Current estimates suggest that BR lost over 3 per cent of its inter-city revenue directly attributable to competition within express coach services.

The government's response to these consequences bears some consideration. It is clear that the public sector bus operators have learned to live with the spirit and purpose of deregulation — but so have those responsible for transport policy learned much from that experience. In this respect actual legislation for stage operations in 1986 embodies most of what has been practised in express coaching and the Trial Areas, but with additions. The government's view is that although competition did take place, it was unnecessarily constrained, both because of the inherent market power of the National Bus Company, and because of the barriers to entry presented by, for example, ownership of bus stations. The Act tackles these problems firstly, by privatising the National Bus Company and, secondly, by placing a duty on those who manage bus stations and facilities prohibiting them from entering into agreements which inhibit competition.

The impact of the Bus Regulatory Reform Act in the USA on inter-urban travel appears to have been quite different from the experience of deregulation in the UK. Although the policy rhetoric sounds remarkably similar — 'The success or failure of a carrier should be determined in the market-place and not in the regulatory arena' — there the similarity ends. Since the passage of the Act it is clear that many new operators have come forward. According to the Interstate Commerce Commission's Office of Transportation Analysis (1984) it is the easing of restrictions on entry which appears to have been the most significant point, combined with major rationalisations of route structures and creation of new services. In the first 18 months of operation, the ICC processed over 2,000 applications for new passenger authority. Over half of these were from new operators. Furthermore, about 12 per cent of all applications were for the operation of scheduled services along regular routes. Given that prior to the

Act effective operational control was in the hands of two companies — Greyhound and Trailways, it is interesting to note that of the new applications for scheduled services, nearly two-thirds were from new independent operators. However, the great bulk — 88 per cent — of applications were for non-scheduled operations, mainly tour, local and special service operations. Applications solely for charter work numbered 1,775 in the same period, of which 54.5 per cent were from new independent operators. The ICC analysis attributes this growth directly to low entry costs associated with the Act.

In terms of new route miles, the Greyhound Corporation estimated that as a result of deregulation, route miles increased by about 7 per cent on top of the 204,000 miles then in existence. Of this, Greyhound and Trailways accounted for 27 per cent, whereas the new independents contributed over 50 per cent.

However, these advances were made at no small cost. The ICC study also examines exits from intercity express services, not in terms of operators leaving the industry but of settlements losing their connections to the intercity bus network. The catalogue charts a steady progress of contraction. For example, prior to the Act one study reported that 1,800 settlements had lost all bus services between 1972 and 1979, implying that over 10 per cent of localities had already no service by the time of the passage of the 1982 Act. Those which lost services tended to be settlements not directly connected to major highway networks connecting the main cities — the most or only profitable routes. By 1981 only 14,600 settlements had a bus service, including most of the country's 436 large urban areas. One estimate suggests that of those areas not served, over 85 per cent had populations of between 1,000 and 2,500 and 15 per cent with populations between 2,500 and 10,000.

Following the 1982 Act, the position seems to have deteriorated further. Within the first 18 months of its operation, operators responding to a survey reported the severing of services to 1,322 locations from their schedules, of which Greyhound contributed over 84 per cent. Of these, 1,045 were in localities which would lose their intercity bus service altogether, of which over a quarter were suburbs or satellites of large central cities. On average the ICC calculated that access in these cases to an intercity route would be about 10 miles away, with connections by local public transport. Therefore just over a million people would be deprived totally of an intercity bus service with no connecting local public transport service.

In what ways is the Act working? Clearly it appears to be working for the Interstate Commerce Commission, but it also, paradoxically, appears to have worked in favour of the two operators who dominated the scene prior to 1982. For example, although the Greyhound's operating revenues fell by around 6 per cent between 1982 and 1983, its net operating income moved from a $3 million loss to a $10 million surplus — even though it carried almost 10 per cent fewer passengers in the same period. This has largely been brought about as a result of what the ICC report calls a more aggressive policy of either dropping loss-making routes or raising fares to profitable levels. These, combined with a fares war with Trailways, 'led Greyhound to actively pursue the downward revision in its unionised wage and benefit structure'. In other words, its management and employees took a cut in their real wages. (It was this aspect which also figured in the British deregulation Trial Area experiment. For example, in an appendix to the *Buses* White Paper, it is stated that 'Recent experience shows that competition enhances the incentive to examine the effect of labour agreements on cost.')

Given the trends within the industry, and the difficulty in raising fares to profitable levels, it is not too pessimistic to argue that the future will be characterised by falling patronage and rising fares. The consequence of that process is clear: a continuance in the severance of localities currently served by bus transport, and further cuts in real wages of those employed in the industry. If that is the future for bus transport in the United States, then it does not seem too implausible to see it translated into the British context. In the meantime, and at the margin, we will perhaps witness a continuing shift out of public passenger transport either to private transport or to a situation where particular trips will not be made.

Railways

The pattern of patronage of the railways corresponds quite closely with the trends at work in public transport in general: a gradual demise in terms of market share in favour of private road transport and air transport. As we saw earlier, however, the history and legacy of the railways has always exerted an incredibly strong influence over the shaping of transport policy. This legacy has its roots set in an era where rail transport offered the solution to long-distance transport, linking industrial centres with markets, and transporting raw materials to stoke up the engines of Empire (in the case of Britain) and to open up new territory (in the cases of North America, Australia, the Indian subcontinent, etc.).

Without going back to these deep historical roots, it is nevertheless important to trace out the pattern which seems to have been present in the development of a rail-based transport policy. The first point is that the railways began as individual operating companies working to commercial criteria of success — profitability, rate of return on investment, etc. The second point concerns the development of the industry's structure — a process of amalgamations, mergers, empires wielding considerable economic and political power. This emergence of monopolies led to direct state interventions and ultimately to the creation of regulatory regimes. The third point stresses the relationships between the regulators and the regulated — a sort of symbiotic relationship developed, whereby as a result of negotiations a mutually satisfactory outcome was the prime requirement.

This very sketchy outline tries to show that in a way governments have been reluctant supporters of rail transport. Apart from maintaining a regulatory environment emphasising safety considerations, operating requirements and pricing policies, transport policy has almost invariably become obsessed with financing rail operations. To do this satisfactorily requires a policy to have a clear objective about what rail transport is supposed to achieve together with measures of performance which it should be required to meet. In practice, no British or North American government had a clear policy objective — apart from an implicit one of minimising the Exchequer's contribution to meeting deficits. For example, in Britain the last major piece of legislation, the 1974 Transport Act, set British Rail the rather vague objective of providing a level of service broadly comparable with what then existed but within a given ceiling of financial support.

In 1947, the railways were nationalised in Britain and the publication of a Modernisation Plan in 1955 was the first real attempt to sort out a role for the industry in the light of changing circumstances. The Beeching Plan in the 1960s continued this logic of modernisation through a process of concentrating on mainline intercity services providing a spine to the national network, closing branch lines, and requiring freight traffic to focus on bulk loads such as coal to power stations. Both of these attempts were, however, concerned with modernising the operational environment. Little attention was paid to the economic environment, or answering questions about why rail transport was important, and what sort of performance targets it could reasonably be expected to meet. However, there was certainly a good deal of weight placed on the railways' financing and debt obligations.

A similar story can be told for North America. In the US, for example, and following the deterioration of railway company finances, the Federal government stepped in and effectively nationalised the US passenger rail network in 1971 — and the US National Railroad Passenger Corporation was the result. Six years later, following a similar catalogue of events, the Canadian government effectively nationalised its passenger railroads — and created VIA Rail Canada Inc.

Instead of state ownership reversing these trends, the level of service provided fell and deficits continued to mount. As Lukasiewics (1984) points out:

> With the exception of the US North East Corridor project, only cosmetic measures were taken to make the trains more attractive. Some new rolling stock was introduced, old equipment was refurbished, ticketing and reservations were streamlined. But, without sufficient capital needed to improve roadbeds and signalling, train speeds could not be increased and rail's competitive position could not be bettered.

That description applies not only to the US, but characterises Britain and Canada also. In Britain one of the earliest responses to this impending crisis was to deregulate the industry from some of its obligations. Both the 1953 and 1962 Transport Acts removed obligations on BR to avoid undue preference. In other words, those within BR management could begin to discriminate in favour of particular activities and thus move away from blanket average pricing policies. It is from this time that it is possible to trace the beginnings of a policy of commercial viability through the railway's own operations and policy decisions, although it was not until 1977 that British Rail could formally operate on a commercial footing.

Arguments about the reform of the regulatory environment within which rail transport operates has been developed furthest, and put into practice, in the United States. Although much of the US passenger rail system was effectively taken into public ownership in 1971, regulation through the Interstate Commerce Commission continued to apply. This consisted mainly of ensuring that the industry did not abuse what was seen to be its monopoly power in particular sectors — especially freight transport. Given its continuing financial plight, Congress passed the '4R's Act' in 1976 — the Railroad Revitalisation and Reform Act. The main aim of this Act was to change the orientation of regulation, away from direct Federal control over rates and towards a freedom of pricing policy where there was no evidence of market dominance. In 1981 the ICC devised four competitive checks on rail pricing policy in terms of the presence or absence of intramodal competition, intermodal competition, geographical competition, and product competition.

In practice this pricing freedom was not automatic, but the change did take the onus of control away from the regulators and towards those responsible for managing the industry. In 1980, the Staggers Rail Act took this freedom a stage further, introducing a more automatic scheme whereby rate approval would be granted subject only to a test of market dominance. In effect the ICC currently regulates only those sectors where there exists either local or national monopolies.

The 1970 Rail Passenger Act, which created the National Railroad Passenger Corporation, also gave federal financial support to unprofitable passenger services. This agency was required to negotiate with the railway companies in an attempt to persuade them to join them in partnership to form Amtrak. By 1971, 18 companies had signed up, in return for either a stock holding in Amtrak or valuable tax deductions.

Within three years, however, another piece of legislation found its way on to the statute book — the Regional Rail Reorganization Act (the '3R's Act'). This was introduced because of the seven major railroad bankruptcies in the north-east states, especially the demise of the once great Penn Central. This Act had two main provisions: firstly, it created the US Railways Association, charged with the planning and financing of rail operations, particularly in the north-east. The USRA in turn set up a private for profit company, Conrail, with the US Department of Transportation holding 85 per cent of its stock. Secondly, it provided further loan guarantees. As a result of this massive intervention, both in terms of a new agency and finance, a significant restructuring of rail operations was undertaken. Conrail's planning function resulted in a new rail network put together from combinations of the seven previous independent companies.

The 4R's Act of 1976 must therefore be set in this context of a developing sense of operational and financial crises. One commentator has described this Act as one which was long sought by the railroad industry, mainly because of its financial package totalling $6.5 billion (US). Besides changing the regulatory environment, the 4R's Act channelled an additional $2.1 billion into Conrail, another $2.4 billion into Amtrak, a further $1.4 billion in the form of loan guarantees for solvent railway companies, $400 million into a national subsidy programme for branch lines, and $200 million into electrification schemes outside the north-east corridor.

What then can be said of the effects of these largely federal initiatives? Firstly, that they represented an attempt to provide conditions of financial stability whilst the industry was restructured. The combination of massive state aid, the creation of two agencies with the powers to bring about this restructuring, and the internalising of the regulatory environment, was the first sustained attempt to provide a breathing space for the railroads' future operations. Secondly, concerted attempts were made to ease the exit barriers which characterised the industry under its previous regulatory regime. So, just as with the Bus Regulatory Reform Act, a major feature of these rail Acts was to improve operators' chances of making physical exits from unprofitable parts of their network. Thirdly, significant improvements were brought about in the area of labour productivity. Again, as a combination of the crisis-ridden state of the industry and its parlous future, labour contracts and agreements were rewritten with the common objective to reduce labour costs. Finally, the overall effect of this complex restructuring appears to have improved the financial health of the industry. Between 1981 and 1984, Conrail has managed to return operating surpluses, so much so that plans are currently afoot to return it wholly to the private sector.

In the meantime operating deficits on passenger operations have been mounting and subsidy payments increasing. Amtrak initially received an operating grant of $30 million; within a year that had increased to over $114 million; by 1981, operating grant stood at

$730 million. This combined with capital grants meant that by 1981 Amtrak had received little short of $5 billion of subsidies in its eleven-year history — the equivalent of 7 cents subsidy per passenger kilometre. However, this is an average for the whole period; the trend has been rising, and so by 1981 the level of subsidy was 11 cents per passenger kilometre, or $42 per revenue passenger.

VIA Rail's performance in Canada is similar. By 1981, total expenditure was $677 million, revenues stood at $161 million, leaving a deficit of $516 million of which $508 million in subsidy was paid. This works out to a subsidy of over $70 per passenger journey. British Rail's performance is equally problematic: in 1980, total costs were of the order of £2,720 million, with revenue running at about £1,920 million — a deficit of £800 million. The 'public service obligation' (a subsidy paid for passenger transport) amounted to an average rate of 2.0 pence per passenger kilometre, or £1.06 per passenger journey.

Given the trends identified here, and the global sums involved, it is perhaps unsurprising that the transport policy response has been to attempt to cut back sharply both the level of services operated and the amount of subsidies involved. In 1980 Amtrak's route mileage was cut, in 1981 fares were increased and services curtailed, and the network envisaged for 1986 implies no passenger trains except in the north-east corridor and a limited number of intercity routes. By 1986, revenue passengers will be required to meet 50 per cent of total costs. As Lukasiewicz (1984) has put it,

In the light of VIA Rail's and Amtrak's experience to date, one can envisage further contraction of traditional passenger rail services as costs escalate and subsidies mount. The faster this process is completed, the sooner the waste of public funds on maintenance of an obsolete technology will be arrested and resources freed for other, more productive uses.

It is surprising but, in fact, Lukasiewicz is a supporter of rail transport. His support stems from contrasting the policy attitude toward the railways in North America with the attitudes in Europe and Japan. He argues that the basic trends at work in North America are present also in these countries, and yet the transport policy response has been quite different. There can be little doubt that this applies to both France and Japan, and also West Germany. Transport policy is qualitatively different, using criteria which are often trivialised by those supporters of narrow allocative efficiency. For example, in Japan the Shinkansen Bullet Train has achieved an operational and financial performance which even the most hard-nosed accountant would admire. These electric trains run on dedicated track at very high frequencies, with speeds up to 210 kmph over the 553 km route. In Canada the average speed of the Montreal to Vancouver train in 1980 was a little over 52 kmph, a figure that has not changed since the 1930s.

The French equivalent to the Bullet Train is SNCF's TGV. Running on purpose-built track, the TGV electric-powered service from Paris to Lyon — a distance of 426 km — is completed in two hours at an average speed of 213 kmph. In 1980, using modern train sets on existing track, the same journey took nearly four hours, even at an average speed of 135 kmph. On present performance the £600 million loan and interest charges will be paid off within 15 years. SNCF, with active government support, has undertaken feasibility studies of new routes, one into the north-west and another to the south-west/Spanish border. Work has already started on the Atlantique route, at a cost of £800 million for the infrastructure alone. The French government's commitment is for 30 per cent of the capital costs. SNCF has calculated the rate of return to be more than 10 per cent. To date the TGV Lyon route has carried more than 34 million passengers since 1981 at an average seat capacity of 70 per cent.

In Britain the policy decision was made not to pursue modernisation based on new dedicated track, but to develop new power units and rolling stock capable of riding on existing track and infrastructure. For 16 years, British Rail devoted a major part of its research and development effort toward the Advanced Passenger Train (APT), aimed at running at 260 kmph with mixed traffic on conventional tracks. A middle range alternative was also developed — the High Speed Train (HST), a more conventional diesel-powered configuration — to fill the void prior to the introduction of the APT. The HST has proved to be the only high-speed train set currently in operation in Britain. The fate of the APT deserves a case study of its own, but briefly major technical problems arose with the tilting mechanism, which was to allow the APT to corner at speeds greater than a conventional train set. British Rail's current plans are to develop a new prototype set — the Electra — with BR committing about £60 million towards its development.

Whether British Rail gets the government's full support to pursue this development is a matter for speculation. In 1983 the Serpell Committee reported on *Railway Finance* and suggested that BR's corporate planning process was in a state of disarray. Serpell's conclusion raised doubts about whether the rather vague objective of the 1974 Railway Act was sustainable.

Since 1975, successive governments had provided subsidy support under the public service obligation (PSO). However, in 1982 British Rail submitted its estimate for PSO of £885 million. The Secretary of State for Transport rejected this bid, offering the Board £804 million. British Rail claimed that they could not meet the 1974 obligation with that level of grant aid; the government's response was that it could through efficiency savings. The result was the setting up of the Serpell Committee. The major problem with Serpell was that it continued in precisely the same vein as most other examinations of rail transport. The emphasis was on questions of interpretations of financial performance, and the treatment by British Rail of the way it allocated its costs. As Nash (1984) points out, 'other options, such as changes in service level and quality over a given network or changes in pricing policy, were given very little attention'. In other words, there remained a refusal to ask the really big questions, not about the railway industry and its control and organisation, but about the role the railway was to play in the future of public transport provision. In the meantime British Rail continues to lurch from crisis to crisis. British Rail estimates that 87 per cent of its costs are associated with track and infrastructure, and do not vary with changes in traffic volume. In this sense cost savings can only be made either by physically reducing capacity or by lowering maintenance standards. For the latter to occur a new set of statutory regulations would have to be prepared by the Department of Transport. For the former to occur, a major review of public transport policy and rail's place in it would be needed. Rowley and Mulley (1983) argue, however, that it is doubtful whether such a review would prove necessary:

> The natural monopoly case for rail network protection does not hold. Issues of sustainability therefore can be dismissed as irrelevant to the transport industry debate, despite the lengthy history of such a belief as a basis for United Kingdom transport policy . . . The railways are the real problem. As a public

corporation accustomed to deficit-financing, and with governments vulnerable to public choice pressures, they represent a continuing inefficient burden to the United Kingdom taxpayer.

A quite different set of arguments has been employed in Sweden which, jointly with Britain, has the the highest level of fares in Western Europe and very low market share. In 1979 the Swedish government decided to relieve Swedish Railways of part of their capital costs. The policy argument was based on an assumption that fares should be more closely aligned to estimates of marginal social costs. This permitted a substantial cut in rail fares of the order of 40 per cent, except at peak times, and generated a large increase in traffic for little additional resource cost. Nash (1984) concludes that, 'Such a policy has much to commend it in Britain, although it would require an abandonment by government of the purely commercial remit'; the indications are that a higher level of investment in BR would yield both social and financial benefits, but that the existing institutional arrangements pose severe problems for the efficient allocation of railway investment resources.

In the meantime the French SNCF began work on the TGV Atlantique route in 1985, with the entire system becoming operational in the autumn of 1990, and only a few months before British Rail completes its east coast mainline electrification programme. Paradoxically, it seems that rail modernisation is also beginning to revive in North America. In 1981 the Japanese National Railways and Amtrak initiated feasibility studies of dedicated track bullet train routes in the US, and four high-density corridors were provisionally identified: Los Angeles to San Diego, Dallas/Fort Worth to Houston, Miami to Orlando to Tampa, and Chicago to Detroit. This is not to say that bullet trains will be developed in Britain. The real import of these developments concerns the way in which complex solutions to complex transport problems are at least being thought through systematically and with an examination of the role of rail transport in the broader jigsaw which constitutes public transport policy.

Air transport

Air transport has been heavily regulated both with respect to quantity licensing and quality controls. It also exhibits all the characteristics normally associated with state intervention, one of the most important being the close relationship between the civil and military wings of the industry. For example, all countries operating both civil and military operations control their airspace through a wholly public agency with complete ownership of their product — airspace — with very rigid entry barriers and fully regulated use rights. The only exceptions apply to general aviation — mainly very small, light aircraft with limited operational characteristics; even so, basic quantity and quality controls apply.

The current regulatory regimes in air transport, with one exception, all have their roots in the Age of Protection. In the US the Civil Aeronautics Act 1938 created the Civil Aeronautics Board (CAB): in the UK it was the Civil Aviation Act 1946, and in Australia it was the Air Navigation Act 1920. Since it is the exception which is supposed to provide the rule, it is interesting to note that the CAB became the first major victim of the Age of Contestability. Under the Airline Deregulation Act 1978, the CAB's life literally came to an end on 31 December 1984.

In terms of changes to the regulatory environment witnessed in recent years, it is perhaps the air transport industry where deregulation has been the most prominent. In the early days, public policy was directed at controlling the development of air transport, on the grounds outlined earlier: firstly, because of the inherent monopolistic tendencies which were perceived to be present; and secondly, because of the safety problems. A third factor, which was particularly significant in the US, was the linkage between passenger air transport and mail contracts (see Panzar, 1983).

Under the 1938 Act the CAB assumed comprehensive controls over entry, exit, route competition, levels of service and pricing policy. The CAB was responsible for issuing certificates of 'public convenience and necessity' to those carriers deemed 'fit, willing and able'. Controls extended to domestic and international trunk routes and carriers, to those providing a local service, to supplemental non-scheduled operators, and to freight forwarders. If CAB controlled the carriers, the regulation of safety, airports and air traffic control was and remains administered through the Federal Aviation Authority (FAA). Besides its certification of air-worthiness procedures, the FAA dispenses federal funds through the Federal Airport and Airways Trust Fund to aid airport developments, and is responsible for the production of a national airport system plan.

Besides its regulatory role, the CAB was also charged with promoting air transport. The main arguments used to support this objective were to avoid what was seen as the inevitable cut-throat competition following the demise of the National Industrial Recovery Act in 1935, and the need to protect the public from localised concentrations of power. This was the stance adopted by the Board well into the 1970s: 'to enable the Board to prevent excessive competiton which, because it is uneconomic, would inevitably drive up fares' and 'competition on all but a few individual city-pair markets is confined to a relatively limited number of carriers – and many markets are served by only one carrier'.

The impetus for this style of regulation was to be found in the excess capacity which characterised the common carrier industries in the Great Depression. Because of these conditions, and through the strong support given to the Board by the carriers, the CAB presided over what Altshuler and Teal (1979) call a system of industry self-government. What this meant in practice was that the Board granted very few new route certificates. By the middle 1950s, about 60 per cent of total revenue passenger miles were on routes served by a single carrier. Such routes were highly profitable and the subject of local monopolies.

The most important events to alter this established pattern of provision radically were the result of technological advance (with the introduction of jet aircraft) and the advent of a deep economic recession (resulting for the first time in a decline in real terms of the number of passengers and amount of freight transported by air). These two events, spreading as they did over a period of years, completely changed the economics of airline operations, and with it began the process of reviewing the regulation of the industry.

Given the financial buoyancy which characterised air transport operators in the 1960s, the CAB was inundated with requests from established carriers to operate on what were seen to be an increasingly profitable core network of routes. The Board ceded to many of these requests, making new route awards in the process. The introduction of this carefully controlled competition resulted, however, not in direct price competition but competition in terms of the level of service offered.

On the basis of what turned out to be optimistic forecasts of airline patronage, the rapid substitution of jet aircraft, and the competition for level of service, the result was a gradual decline in the average load factor — the proportion of aircraft seats filled by fare-paying passengers — from 65 per cent in the middle 1950s, to an average of 54 per cent during the 1960s, declining to 49 per cent by 1970. Not only did the load factor decline but the average airline fare was also reduced by 14 per cent, even though revenue passenger miles had increased at a rate of over 14 per cent annually. So long as growth was rapid and profits high, neither the Board nor the airlines paid much attention to the combined effect these trends were having on the economics of their operations.

The general consensus between the carriers and the CAB was shattered by the middle 1970s. Several factors combined to threaten the regulatory environment. The first was the onset of recession and the fact that between 1969 and 1970, patronage fell in absolute terms, and the previous very high rate of growth slowed down; secondly was the decision reached in the Appeals Court that discriminatory pricing strategies were unjustly discriminating; thirdly, many carriers were in deep financial trouble, largely because they had placed firm orders for the new widebodied jet aircraft; and finally, because of a new mood which was beginning to seep into public policy in general that regulation produced very few benefits. The advent of the Yom Kippur war in the Middle East, combined with the Arab oil embargo, intensified these events.

The Board changed its overall direction in the light of these events, but intervened to curtail level of service competition, suppressed the practice of discount fares, and attempted to reduce the excess capacity which was beginning to appear among the carriers. The policy objective behind this strategy was quite clear: to safeguard the financial health and market shares of the established operators.

The impetus for price competition came from neither the Board nor the carriers. In 1975, the Federal administration submitted a proposed Civil Aviation Act, whose centrepiece was the introduction of price competition and a gradual relaxation of some of the CAB's regulations. Although this legislation was not enacted, the scene was set for a further change in direction for the Board with the appointment of a new chairman who supported a relaxation of its regulatory stance. A series of small changes in the armoury of CAB's regulations, particularly with respect to allowing discount fares and a more flexible fares policy, coincided with a dramatic change in the state of the airline industry: in just one year, 1977, traffic had grown by 8 per cent and profits by 82 per cent. The momentum for further rounds of regulatory relaxations was gaining considerable support, with the result that air freight became fully deregulated in 1977 and air passenger deregulation was well under way by 1978.

By 1978, the CAB itself had engineered a number of deregulatory devices. Although controls on new entrants remained extensive, except in commuter air services, the main thrust of the changes was to introduce a limited form of price-based competition together with easing restrictions on, for example, exits from unremunerative routes and encouraging mergers. The 1978 Deregulation Act took these steps one stage further, directing the CAB to continue the process already in train but adding a number of explicit policy changes, for example, to place 'maximum reliance' on competition when making policy decisions. The final nail in the CAB's coffin was provided by a clause in the Act which specified that the Board itself would cease to exist by the end of 1984.

In 1977, the General Accounting Office of the Federal administration arrived at an estimate of the cost of air transport regulation of between $1.4 billion and $1.8 billion. But the demise of the CAB has not meant a cost saving of that magnitude, mainly because of the provisions contained in the Deregulation Act for maintaining quality controls through the FAA and the financial programme to help overcome what were seen to be difficulties in the transition period to a deregulated environment. To ease the transition, Congress voted for the Essential Air Service Program (EAS). The main purpose of EAS is to provide subsidies for air services from small towns to larger towns. A major fear of the Deregulation Act was seen to be the probable demise of peripheral air services on less profitable routes in favour of those parts of the core network which were capable of earning higher returns.

Just as in the case of the bus and rail industries, air deregulation was aimed at, firstly, shifting the onus of responsibility for service provision away from the regulatory body and towards the carriers, and secondly, providing the conditions or breathing space whereby the industry itself could be restructured and reorganised in the face of a changing economic environment. Under the 1978 Act, the CAB was charged with relaxing and shedding controls over entry, route competition, levels of service, pricing policies, and exits built up since 1938. Of these controls, it was the possible effects of route and service exits upon peripheral localities which created the greatest unease and hence led to the EAS programme. The US Transportation Research Board in 1973 estimated that by 1988, when the subsidy was finally to be phased out, between 15 and 20 per cent of communities presently served by air under the EAS would have built up traffic levels which would cover the costs of operation. A further 30 per cent, it was estimated, would lose their air service, but were within reasonable driving distance of a hub airport. The remaining 50 per cent of communities would in effect most probably lose their air services altogether. As Higdon (1985) points out, this means that about 70 cities and towns receiving subsidy would effectively be cut off from access to air transport services.

Since 1978 the effect on carriers has been influenced by both the continued and steady growth in patronage and profitability, and by their own new found ability to reorganise their corporate strategies. The combination of deregulation and recession has clearly strengthened the industry's ability to restructure, reflected in the number of mergers and rationalisations between smaller carriers and larger operators, together with consolidation of routes and services. An example of the sort of cost savings anticipated as a result of this sort of merger is reflected in the new employment contracts negotiated between managements and unions. Northwest Airlines' agreement with the International Association of Machinists, for example, features lower pay scales for new employees at between 20 and 50 per cent below prevailing wage rates for the first six years of employment, whereupon the higher wage rate will apply. Coupled with an agreement to increase job flexibility, any new mechanic taken on will receive 21 per cent less than an equivalent established mechanic; for aircraft cleaners, the difference is of the order of 45 per cent.

Considerable space has been devoted to the US experience of regulation, largely because it is the only country where regulatory policy has been stood on its head, but also because it might suggest the sort of broad impacts which are likely to result from other countries pursuing similar ends. In the UK, the traditional approach to air transport regulation has been very similar to the US, as indeed has Canada's, Australia's and New Zealand's.

In South Australia, for example, controls — particularly pricing — are operated through the Independent Air Fares Committee (IAFC) set up under the Independent Air Fares Committee

Act of 1981. As Starkie and Starrs (1984) note, the IAFC has been preoccupied with fares policy and the setting up of individual route fare levels to ensure that the level of air fares is closely related to the cost of providing the services. Approval of fares involves examining fare applications, comparing the proposed fare with the distances involved, and drawing attention to major departures from industry-wide standards. New entrants are permitted, but if the fare is lower than the established carrier then approval must be sought. This solution reflects the sort of relaxations made in air transport regulation following the Federal decision of 1979 to pursue a more liberal policy. The principal form of competition, however, remains service rather than price based. Starkie and Starrs's assessment of these relaxations suggests that there has been a significant increase in both services offered and their frequency, with the development of new routes often connecting centres for the first time. There is little evidence of hit-and-run entry.

In the UK, the responsibility for regulation has been vested in a single agency. At present the CAA has complete authority over all aspects of air transport, both with respect to quantity and quality controls. It has the power and discretion to grant, revoke or seek modifications to its operating licences, according to the traditional criteria associated with regulation.

The CAA's licensing system has a dual function: on the one hand it regulates the industry, but on the other it is charged with promoting air transport. Its objectives are: to ensure that routes are properly serviced; to avoid overcapacity in the industry; and to further the development of British civil aviation. Although the 1946 Civil Aviation Act created the regulatory regime, prior to 1971 (and the creation of the CAA) responsibilities were vested in a wide range of largely different central government departments. The first major review of air transport regulation was in 1969 following a report by the Edwards Committee which had been charged with examining the whole range of licensing systems and competition policy. The report was based on two major assumptions: firstly, it foresaw a trend in mergers and rationalisations among smaller independent carriers; and secondly, an equivalent trend in the consolidation of national state airlines to form a single 'first force' carrier. The review of competition policy took place against the background of the activities of the Air Transport Licensing Board, the body created in 1960 to control entry and fares. As Rowley and Mulley (1983) comment:

Despite a lengthy list of criteria provided by the Act, it was clearly the intention of the Government to leave the Board substantially unfettered as to its interpretation of its general duty. Despite such flexibility, however, the Board attempted to create clear precedents by publishing the reasons for its decisions and by summarizing in its annual reports what had been the important principles enunciated and applied during the preceding year.

The consequences of this strategy were: firstly, it created dissent between the Board and air carriers, but secondly, it allowed the automatic right of any party to a particular licence application to appeal directly to the Minister. On the one hand the ATLB appeared to be making its own air transport policy, and on the other, it had become subjugated to the Minister — a situation which the Edwards Committee was expected to resolve. The Civil Aviation Act of 1971 disbanded the ATLB and replaced it with the CAA, providing it with more comprehensive regulatory controls, but with a clearer set of policy guidelines.

The 1971 Act attempted to strike a balance between the cost advantages likely to be reaped from larger scale operations, the efficiency gains accruing from service-based competition, and externalities arising from environmental effects. The Act stated that the CAA should grant operator licences to more than one carrier on any given route, so long as the traffic was likely to support profitable services and the level of service was likely to be improved.

The result of this reorientation, coupled with limited relaxations of competition rules, meant that the CAA became more active in its promotional role of encouraging mergers, promoting service competition and protecting the established nationalised monopoly carriers. On the international scene, British Caledonian was formed as a national carrier from two independents immediately prior to the CAA's new role. As a result the CAA encouraged BCal as a 'second force' carrier in competition with the two state-owned national airlines — BEA and BOAC. On the domestic front Air UK was formed from a merger of five small independent carriers and British Midland Airways was also encouraged to expand its operations as a 'third force' airline. The CAA was also partly responsible for promoting the merger between the two state-owned airlines to form British Airways in 1972.

More recently the CAA has been pursuing an increasingly liberal practice within the policy constraints imposed upon it under the 1971

Act. In 1984 it published its own review of airline competition policy, and made it clear that it recognised the need to review its operations. Its attitude to competition is illuminating:

> The Authority sees competition as a valuable mechanism for securing in the right circumstances the objectives laid down in the Act and not as an end in itself . . . Short term gains in user satisfaction must be carefully weighed against the longer term need for the sound development of an efficient and competitive industry on which the continuing satisfaction of user demands must depend (CAA, 1984a).

It is perhaps interesting to note that this more cautious line of argument followed a thorough review of the experience of US style deregulation (CAA, 1984b).

The CAA has also published proposals to relax its control of domestic air fares on all routes where its regulatory powers were in force. Under its 1971 Act the Authority has the power to implement such a change in the form of new or revised rules. In this case the new rules require carriers to notify the CAA ten days before adopting a new fare. The understanding is that the Authority will only intervene if it sees the new tariffs as either predatory or the result of monopoly behaviour. However, as it has declared, 'such intervention would be expected to be very rare'.

The Department of Transport (1985) has also published a policy statement on air transport and airports. An original proposal in 1984 to allocate arrival and departure slots by auction has disappeared, and in its place is a new regulatory responsibility for the CAA. The current intention is to control air traffic through a traffic distribution policy, with the CAA having the duty of developing and promulgating a scheme for allocating scarce airport capacity. The new policy guidelines offered to the CAA strengthen rather than weaken what is turning out to be a revamped form of regulation, which as yet has not struck a similar chord in regulating either the bus or rail industries. This new, more cautious policy orientation appears to fly in the face of the policy changes proposed for other domestic transport modes in the UK. It is also quite different from the more recent advances in deregulation made in North America, and from those calls for liberalisation being pursued in the Council of Europe.

What this account of air transport regulation has tried to

demonstrate is the almost irresistible force for deregulation, led by the US, which has been at the centre of policy debates in recent years. As Altshuler and Teal (1979) comment, 'This procompetitive triumph in the field of air transport regulation is unprecedented in the annals of American transportation.'

Concluding Comments

This rather lengthy treatment of the regulation and control of transport reflects its increasing priority in terms of the development of future public policy for transport. We have seen that the previous, accepted approach to ownership policy based on regulation has become increasingly questioned. The traditional stance, based on controlling entry, exit, competition and pricing, both with respect to quantity and quality, has become out of step with the reassertion of arguments based on the unquestioned dominance of competition and allocative efficiency. We have also seen that this neo-liberal approach has been born out of a period of intense and sustained economic crisis. Previous policies, based on attempts at improving the allocation of transport resources through a state-based administrative process of planning and co-ordination, have broken down. New policies have been introduced aimed specifically at improving the conditions whereby productive capacity can be reorganised within the private sector of the transport industry with the minimum public intervention. This has been coupled with the return of state-owned and managed operations to the discipline of the private sector market-place and to the transfer of public assets back into private ownership. The main thrust of this new approach follows the banner of deregulation.

Whether and for how long this new found fervour for liberalisation in favour of the market-place continues is a matter for pure speculation. However, and as we have seen particularly in the UK examples, deregulation policies have been pursued to date largely with more rhetoric than substance. The style and form of regulation has changed and arguably will continue to change for the foreseeable future. But there are voices beginning to be raised about the consequences of deregulation which may suggest modifications to what at first sight may appear as a wholesale dismantling of regulatory controls. Much depends on views about three largely speculative future developments. Firstly, when, or perhaps if, the present deep economic recession disappears, and there is a return to some measure of general stability in world trade and economic growth, then the conditions which created the need for regulation in the first place may reappear, although probably in a different form. Following the massive reorganisation currently taking place in the ownership and control of transport operations, then it is possible to argue a case that monopolistic or other forms of what is generally seen as undesirable competition may return. Whether this would mean a new and heightened role for those existing regulatory bodies responsible for antitrust and fair trade policies is a matter for debate.

Secondly, and following from this latter point, will the present established operators in the transport industry cede to such relaxations, or will they have the corporate political power to influence the pace and scope of liberalisation, and ultimately turn such changes to their own advantage? It is more than feasible to argue that the benefits of the limited forms of deregulation in the bus industry have accrued largely to the established operators, particularly in their ability to turn round their previously parlous financial circumstances, and in the process renegotiate labour and productivity agreements. This also appears to be a major benefit in the airline industry, particularly in the US.

Finally, the intitial impetus for deregulation — or ultra-free competition, as it is perhaps better called — coming as it did from neo-liberal academic economists, is also beginning to be questioned. The main points of contention have been argued by Shepherd (1984):

> Implausible assumptions have been applied on an abstract plane to reach not only 'insights', but also emphatic conclusions and wide policy lessons. The system hangs in the air, lacking a foundation or even plausibility.

In his account of the experience of airline deregulation as an example of the need to create conditions for ultra-free competition (that is, complete freedom to enter into and exit from a market), Shepherd argues that 'Under deregulation during 1975–84, the industry has displayed rising competition and more flexible pricing. But this does not prove the validity or generality of ultra-free entry theory.' One important reason for this being so is because airline markets are not well defined; it therefore has to follow that the roles of barriers to and conditons of entry and exit are equally unclear. He concludes with a view that wise public policy choices should remain based on the evidence of the accumulation of past research, with its focus on actual competition, rather than rely on a theory whose basic assumptions appear increasingly questionable.

References

Altshuler, A. and R. Teal (1979), 'The Political Economy of Airline Deregulation' in A. Altshuler (ed.), *Current Issues in Transportation Policy* (Lexington Books, Lexington).

CAA (1984a) 'Airline Competition Policy', *CAP 500* (Civil Aviation Authority, London).

CAA (1984b), 'Deregulation of Air Transport', *CAA Paper 84009* (Civil Aviation Authority, London).

Department of Transport (1984), *Buses*, Cmnd 9300 (HMSO, London).

Department of Transport (1985), *Airports Policy*, Cmnd 9542 (HMSO, London).

ECMT (1985), *The Evaluation of Past and Future Transport Policy Measures; 10th International Symposium on Theory and Practice in Transport Economics* (ECMT, Berlin).

Evans, A. and L. Hoyes (1984) 'Bus Services in the Hereford Trial Area', *School for Advanced Urban Studies WP44* (University of Bristol, Bristol).

Fairhead, R. and R. Balcombe (1984), *Deregulation of Bus Services in the Trial Areas*, Transport and Road Research Laboratory, Report 1131, Crowthorne, England.

Glaister, S. and C. Mulley (1983), *Public Control of the British Bus Industry* (Gower, Aldershot).

Hibbs, J. (1982), *Transport Without Politics . . .?* Hobart Paper 95 (Institute for Economic Affairs, London).

Higdon, D. (1985), 'DOT Seeks End to EAS', *Air Transport World*, April, pp. 74–5.

ICC (1984), *The Intercity Bus Industry* (Office of Transportation Analysis, Interstate Commerce Commission, Washington D.C.).

Lukasiewicz, J. (1984), 'Passenger Rail in North America in the Light of Developments in Western Europe and Japan', *Transportation Planning and Technology*, vol. 9, pp. 247–59.

Nash, C. (1982), *Economics of Public Transport* (Longman, London).

Nash, C. (1984), 'Rail Policy in Britain — What Next?' *Transportation*, vol. 12, pp. 243–59.

Panzar, J. (1983), 'Regulatory Theory and the US Airline Experience', *Journal of Institutional and Theoretical Economics*, vol. 139, pp. 490–505.

Rowley, C. and C. Mulley (1983), 'Transport Regulation in the United Kingdom', *International Journal of Transport Economics*, vol. 10, pp. 443–80.

Self, P. (1975), *Econocrats and the Policy Process* (Macmillan, Basingstoke).

Shepherd, W. (1984), '"Contestability" vs. Competition', *American Economic Review*, vol. 74, pp. 572–87.

Starkie, D. and M. Starrs (1984), 'Contestability and Sustainability in Regional Airline Markets', *Economic Record*, vol. 60, pp. 274–83.

Taebel, D. and J. Cornehls (1977), *The Political Economy of Urban Transportation* (Kennikat Press, London).

Webster, F. W. (1985), 'Passenger Transport: Investment and Financial Decisions in Urban Areas' in 'The Evaluation of Past and Future Transport Policy Measures', *10th International Symposium on Theory and Practice in Transport Economics* (ECMT, Berlin).

5 INTERNATIONAL REGULATION AND CO-ORDINATION

National Policies and International Transport

The full exploitation of a nation's comparative advantage has long been recognised as an important component of maximising national economic growth. Going back to the classical economists of the early nineteenth century (and most notably Ricardo) the importance of *laissez-faire* has been emphasised so that the full benefits of national specialisation in production may be reaped. Countries with a comparative advantage in the production of manufactures would specialise in their production while those with a comparative advantage in agriculture would focus on production in this sector. Trade between the countries would be to the benefit of both and the world's overall productive capacity would be maximised.

The trouble with applying this philosophy in the real world has always been that while *laissez-faire* maximises aggregate world output it need not be to the specific advantage of any individual country. Where a nation has an absolute advantage in production it may be advantageous for it to exercise monopoly power and, by restricting output of some goods and so forcing up their internationally traded price, increase its well-being at the expense of other states. Equally, individual monopsony buyers of goods may benefit by restricting imports from other nations and thus gain from lower prices at their expense. Over the years a succession of agreements (e.g. the General Agreement on Tariffs and Trade — GATT) and trading unions (e.g. COMECON and the European Economic Community) have been developed which have attempted to remove or at least reduce the effects of various tariff, quota and subsidy systems which have been introduced at various times by national governments in furtherance of their individual economic well-being.

What has this discussion of international tariff reform to do with the issue of transport policy? The simple fact is that a national government can manipulate its transport policy to produce the same effect, or at least one very similar, to that of a tariff on imports or a subsidy for its exports. By regulation and control it can influence the costs of transport services so that the prices paid domestically for imported goods or by foreign customers for its exports are distorted in a manner identical to that of direct fiscal intervention in international trading markets.

The ways in which government can influence international transport costs are numerous. In most countries it is the national government which provides the major items of infrastructure — the ports, airports, railway track etc. Provision in itself gives power because by its very location or capacity the infrastructure can influence the magnitude of a country's trade. Railway networks can be constructed to facilitate the rapid and cheap export of goods but provide only scant access to areas of the country offering the greatest potential markets to importers. Ports can be designed to handle export goods easily but offer little capacity capable of dealing with the commodities most likely to be imported. Access to the infrastructure is another domain where discretion can be exercised. Quotas can be imposed on the number of foreign freight vehicles allowed on a country's road network, forcing importers to employ domestic hauliers (which, incidentally, may be a distortion to trade in itself if this results in the artificial exportation of haulage services). France, Germany and Italy have traditionally done this with road haulage activities. Technical restrictions can also favour national carriers and may force economically costly transhipments at borders if only domestic road vehicles meet capacity or safety standards or if the railway system is of a unique gauge. The UK, for example, has often faced complaints in this respect because of its relatively low maximum weight for lorries.

The power to charge for the use of the national transport infrastructure offers a further device by which a government may attempt to influence trade. Landing fees at airports (which are normally weight-based), for instance, can be so tiered that the costs of importing air freight exceed those of exporting. Alternatively, they may be so designed that tourist traffic is attracted at the expense of alternative and, in strict economic terms, cheaper facilities located in competing centres. The charges for the use of road and rail track may be so arranged that exporters receive hidden subsidies. If coal or iron ore deposits, for example, are located near border areas, carriage rates may be highly tapered to encourage their export to nearby foreign markets — a system, for example, employed by West Germany in the post-Second World War period so as to penetrate French and Benelux markets — and protect their own industries. Alternatively, by levying high temporal fees (e.g. an annual licence fee) but lower direct usage fees (e.g. fuel taxes) a country can seriously push up the costs confronting foreign transporters conducting business into its domain. Because relatively little of their carriage is done in the country the average cost per mile to a foreign operator is high — a complaint the Canadian road hauliers made against proposed vehicular tax reforms in the USA during the early 1980s.

More direct measures are also not unknown and while these are often introduced for reasons of military or colonial necessity (e.g. the UK's Navigation Codes initiated in the seventeenth century) they also act to influence the pattern of trade and to violate principles of *laissez-faire*. These direct measures themselves may be of several different types. They may restrict the carriage of traded commodities to specific modes of transport (usually a national carrier) or force the use of specific pieces of infrastructure (as, for example, the French did in 1982 by limiting the importing of video equipment to certain ports and routes).

Institutional arrangements may also serve to influence the cost of international transport. Normally in domestic operations vehicles and drivers are only subjected to periodic or random inspections of their documentation but crossing national boundaries generally involves lengthy and costly delays as considerable paperwork is undertaken (enough, indeed, on the French–Italian border to bring the French road haulage industry out on strike in 1983). Again this serves to protect local industry, particularly manufacturing and agriculture, from the rigours of equitable competition from outside.

Government may also intervene, again in numerous different ways, to protect its own, domestic transport industries from external competition. Cabotage, in particular, is often viewed in the same way as the 'dumping' of goods in a market and is the subject of particularly severe restrictions. Many actions in this area often involve policies closely entwined with those whose primary objective is to assist national manufacturing but because, in some cases, transport services are themselves important generators of jobs domestically or may form a major export industry in their own right there are also instances where the aim is more specific to transport *per se*. In other countries the production of transport products (and here we must include the supply of consultancy and civil engineering expertise as well as that of cars, lorries, aircraft, ships and other vehicle manufactures) is deemed important (often for reasons of domestic economic stability and the retention of traditional sectors rather than on grounds of strict domestic economic efficiency) and government intervention limits the workings of the free international market. Not only, for instance, are domestic car industries protected by design standards, quota systems and tariff regimes but also national professional bodies impede the workings of the international market in labour expertise.

Just as national government policies may be geared to intervene directly in international transport so in some instances the lack of effective or co-ordinated action may equally lead to impediments in the efficiency of the market. The supranational power of transport cartels in some sectors transcends national policy and, by giving transport suppliers monopoly powers over shippers, acts to reduce the volume of trade. Clearly few of these cartels confess to such actions and justify their collusion in terms of reduced uncertainties or social provision but irrespective of this there is no doubt that the outcome represents a deviation from the *laissez-faire* position (see, for example the section on Shipping conferences, later in this chapter (pp. 135–40)). Whether this is desirable or not is, of course, dependent upon specific circumstances, but lack of co-ordinated and agreed international policies, despite frequent complaints, discussions and enquiries, suggest that the actual situation seldom conforms with that which many national governments would prefer.

The multifarious national interventions in the international transport market which occur have two distinct effects. Firstly, many national policies are adopted to alleviate some short-term problem or in an attempt to gain some commercial advantage. In general, they bring forth retaliatory actions from trading partners or from alternative, national suppliers. But even when this is not the case, the long-term effect is often a reduction in the efficiency of the international economy. If the policies are all designed to protect local industry or domestic transport suppliers then this means that almost inevitably the cost of transport is pushed artificially high and overall production is kept below its maximum potential. If the policies involve transport subsidies, say to aid exporting firms, then the long-term impact is equally damaging. The overall cost of transport is too *low*, meaning that excessive resources will be tied up providing facilities to meet the high level of demand for transport services. Those which are normally deemed inefficient because of their exorbitant transport needs now become competitive with their more economically efficient rivals.

While the overall effect of independent national policies in the international transport field is almost certain to constrain aggregate international economic expansion there will also be an overlapping trade diversion effect. Certain countries (or specific sectors within them) are likely to have their growth rate curtailed quite significantly. Empirical evidence suggests that many Third World countries suffer disproportionately in this way. Such countries are highly dependent for their economic growth on the sparse foreign earnings they obtain from their often limited range of exports. In terms of market power, their scope for bargaining or delaying shipments is limited. Additionally, these countries seldom have their own international transport industry but depend heavily upon the fleets and facilities of developed countries. When it comes to the initiation of national transport policies to protect their own interests, therefore, it is the Third World nations which tend to bear the brunt of the impact of slower worldwide economic growth. The unfavourable distributional impact on less developed economies is even more pronounced when one realises that most tightening up of national policies occurs when the world economy is in recession. In order to preserve their own income and employment at times of economic contraction, developed economies tend to become more protectionist. Third World countries have few enough reserves to tide them over such periods and limited means of initiating effective countermeasures.

Trends in International Transport

International transport has always been very closely tied to the state of the international economy. When there is general prosperity then the demand for both finished and primary goods puts pressure on suppliers and stimulates effort to shift goods between markets as the highest price is sought. This relationship, of course, holds more generally and applies equally to domestic markets for transport services. There are a number of factors, however, which suggest that over the next decade or so the demand for international transport services is likely to rise much more rapidly than domestic demands in most countries. Some of these factors relate to the accelerating trend towards greater intra-industry specialisation within the manufacturing sector while others reflect changing tastes and improved communications.

Traditionally, it has always been the perceived wisdom that the majority of international trade involves the movement of raw materials from mainly Third World countries to industrial nations with a flow of manufactures in return. In fact, even this simple picture has, in many ways, always been somewhat inaccurate. Many of the industrial nations themselves have long been net exporters of primary products while, particularly in recent years, many of the Third World countries have built up export industries in a number of manufacturing sectors. The situation is further complicated by the serious imbalances which have in the past existed both in terms of the value of the goods shipped (generally manufactures having a much higher value per ton) and the nature of the transport services required (manufactures often being in consignments which do not, for the appropriate mode, constitute a full load). Again this has been very gradually changing with many Third World countries, although still often highly reliant upon primary sectors of production, attracting multinational companies by their cheap labour and low-tax regimes to establish component part and sub-assembly plants within their borders. Labour costs in particular have superseded the availability of raw materials as a major influence in industrial location. Many of the components produced are high-value, low-volume items which are then sent (increasingly by air) to Western countries to be combined with locally produced, but bulky components to provide final products for the domestic market (see Helleiner, 1973).

Essentially, although older patterns still persist, trade between the developed and less-developed worlds is slowly evolving along lines which are not dissimilar to those firmly established, if not yet always fully exploited, in trade between the developed nations.

The developed economies themselves are tending to expand intra-industrial trade — trade in goods of a similar industrial classification — with each other to permit a fuller exploitation of economies of scale, scope and experience. While the exact measurement of the magnitude of the swing towards intra-industry trade is highly sensitive to the level of aggregation employed in industrial classification, some general impression is gained from the indices shown in Table 5.1. Here if all trade for each of the countries represented were between sectors (say an exclusive export trade in manufactures and an exclusive import trade of primary products) the index would be zero. As is seen in the vast majority of cases the index is well above zero (the extreme value for exclusive intra-industry trade being 100) and rose appreciably between 1964 and 1977 (a feature even more marked when the indices are compared to longer term historical trends).

Table 5.1: An Inter-Country Comparison of Average Levels of Intra-Industry Trade in Manufactures. Index for complete inter-industry trade is zero

	1964	1977
France	73	76
Italy	60	58
Netherlands	67	69
UK	64	72
Germany	53	64
Austria	53	63
Sweden	58	63
Norway	39	40
Switzerland	55	58

Source: Z. Drabek and D. Greenaway (1984), 'Economic Integration and Intra-Industry Trade: The EEC and CMEA Compared', paper presented at the 15th Atlantic Economic Society Conference (Paris).

These changes have implications for the current and future demands which are going to be put on the international transport industry. Before looking at these, however, it is perhaps helpful to offer just one illustration of the direction in which the trend is moving; namely the international, mass-production car manufacturing industry.

Until comparatively recently, cars were exported complete, following assembly, or in kit form with the vast majority of components supplied to be put together by local labour. A number of factors changed this in the late 1960s, not least the desire of national governments to be seen to have their own car industry and concern about the balance of payments implications of importing foreign vehicles. Where nations had established but uncompetitive car industries, there was also a desire to protect domestic employment. Consequently, barriers were erected to reduce vehicle imports — many were direct quota systems (often of a 'voluntary' nature) or involved required domestic input ratios but others focused on introducing national technical standards which differed from those of the major manufacturing nations. The manufacturers responded by directly investing in the car industries of such countries and thus circumventing many of the restrictions. Production of an entire range of individual cars for each market clearly did not, however, permit the full potential of large-scale manufacturing to be reaped. The largest manufacturers, therefore, initially opted for specialisation with one or two models in their ranges being produced in each country and then shipped, normally complete, to the various other markets. While producing some cost savings, once more the full benefits of scale economies were still not exhausted under such a regime. More recently, it is components which have become more mobile, often being both interchangeable between models (and thus minimising production costs) and capable of being combined with locally produced components to provide a product 'tailored' to specific national tastes or technical specifications.

Some idea of the level to which automobiles have become internationalised is given by the crude breakdown seen in Table 5.2 of the national origins of the components of a small family car on sale in the UK in 1984. (In fact, even this is not complete since many items which go into making up the major components we have listed do themselves contain imported parts.) The reason General Motors finds this particular combination of components cost-effective is as much to do with national policies of the supplying countries as with the forces we associate with conventional economics. (It is worthwhile, for example, to ship engines from Australia because this permits General Motors to exploit a loophole in Australian trade regulations it to fit Japanese components to cars it produces and sells in Australia.) The mix is, therefore, likely to change over time not simply in response to free market economic forces but also as a reaction to national commercial and industrial policies.

The key point to be made is that there is a shift towards the international transport of components rather than final products and this, in its turn, has important ramifications for the transport sector.

Firstly, the manufacturers are essentially reaping the benefits of large-scale production at the expense of greater demands for

Table 5.2: Composition of Components for Vauxhall 1984-Model 1.6-Litre Manual Gearbox Astra [1.3 version given in parenthesis]. By country of origin

Australia	Engine
Austria	[Engine, gearbox]
Eire	Electrical wiring
France	Radiator
Japan	Gearbox
United Kingdom	Seat belts, wheels, front doors, floor pan, steering, bonnet, front wings, instruments, interior trim, upholstery, glass
West Germany	Tailgate, rear doors, suspension, fuel tanks, bumpers, headlamps, seat frames, roof

transport imports. Therefore, even if total production were to remain constant, transport requirements would rise as component specialisation becomes more extensive. Production, given both general upward trends in income, but also the lower real cost of manufacture resulting from this international division of production, is, however, also likely to increase, resulting in even greater demands for international freight transport services. While forecasting, especially of short-term trends, is notoriously difficult in the international transport sector, a number of key agencies are,

Figure 5.1: Forecasts of the Growth Rates of Domestic and International Freight Traffic in the EEC

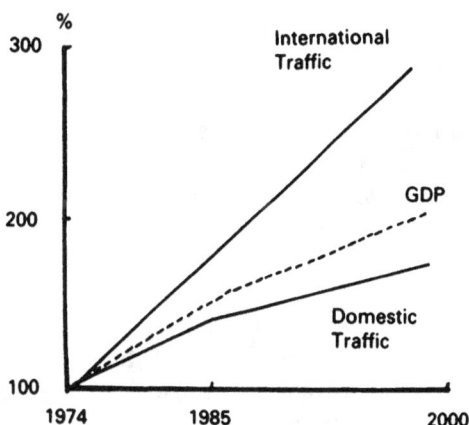

Source: EEC Freight Forecasting Study, 1979.

for these sorts of reasons, suggesting that the future growth in international traffic is likely to exceed that of national, domestic transport. This, for example, is the situation envisaged by the EEC (see Figure 5.1).

Secondly, the demands placed on the international transport system in the future are going to be somewhat different to those of the past. Components are less bulky than final products and are likely, as a generalisation, to have a higher value – weight ratio. This means that the comparative advantage of transport modes best suited to bulk commodities is likely to be further eroded. International specialisation is also likely to put a premium on the speed of service which is offered and, particularly, on the re-liability of that service. The geographical separation of component manufacture from final assembly means that any bottleneck in transport arrangements curtails assembly and disrupts the pro-duction process. To hold large inventories of components in order to hedge against any such uncertainties involves tying up scarce resources. Equally, if the transport system can provide only a slow service this means that substantial stocks of components are perpetually tied up in transit — once again this is extremely costly.

Thirdly, many of the components used in modern, high tech-nology industries are extremely fragile and valuable. Further, the rate of technical progress means that many manufacturers rely upon essentially short-term advantages in the technology of their products — to retain this for as long as possible they seek security in their transport and particularly that of key components. Greater personal or corporate control over transport at the international level is, therefore, likely to be a further feature of the evolving situation; a trend essentially mirroring that which has been seen in domestic freight transport for a number of years.

Finally, component specialisation in the high-technology sector means that the location of production facilities may be less in-fluenced in the future by factors such as available land and more by high quality *international* transport facilities. To some extent this trend is already being observed in the USA where ready access to air transport is a key consideration for modern industry. Equally, it is no accident that Heathrow Airport is, in terms of value, the largest international port in the UK with considerable developments of high-technology industry taking place adjacent to the main transport arteries offering access to it.

While the recent growth in the internationalisation of production is likely to continue for much of the period we are concerned with, there are fragments of evidence (e.g. from West Germany, the USA and some areas of the UK — such as the electronics industry of South Wales) that even longer term trends may produce some limited reversal of this process. While differential labour costs have often been the driving force determining the siting of productive capacity, advances in process technology in many technological sectors reduce labour inputs to such an extent that the advantages of spreading component production across countries is significantly reduced. There is, therefore, a longer term trend back towards geographical concentration of the overall production activity. This is unlikely to be of major importance for international transport, however, during the next decade or more.

It is not just freight transport where changing conditions are likely to necessitate modifications to the way policy-makers view international transport. There are also important changes on the passenger side which are gradually emerging. (In many ways these are more difficult to forecast than for freight, as was illustrated by the development of Concorde in a mistaken anticipation of steeply rising demands for *rapid*, long-distance international business travel.) Two important trends are emerging.

In terms of business travel, while demand is likely to rise as an adjunct of the internationalisation of many manufacturing and commercial activities, the rise is likely to be tempered by advances in information technology. The area where transport demand is likely to rise most rapidly is at the top end of the market where 'face-to-face' contact is still deemed important (although whether this is a function of negotiating efficiency, social cohesion or simple tradition is another matter). The very rapid increase in the number of small airlines catering for executive and business travel (some 215 are already operating in the USA and over 30 in Europe) and the development of high-speed executive aircraft provides some general evidence of the expanding demand for long-distance travel in this market.

The market for international leisure travel has grown considerably over recent years as incomes have risen and work patterns have allowed longer, uninterrupted periods of leisure to be enjoyed. Although there seems to be an emerging consensus that incomes in the developed world will not rise in future at the same pace as in the 1960s and early 1970s, higher incomes are still anticipated. Equally, advances in technology suggest that while people may need to work more hours per day to maximise the potential of recent advances both in the production technologies of manufacturing and the information aspects of service industries, they will have more 'lumps' of time free for relaxation. These trends, coupled with the potential attractions of more distant locations are likely to stimulate demand for longer distance international leisure travel. Essentially time spent on travel will represent a smaller fraction of the overall holiday and thus increase the relative attractiveness of far off places. The exact locations which will benefit, however, are difficult to forecast and, given the uncertainties of floating exchange rates, political instabilities etc. are unlikely to be as directly influenced by international transport policies as by other factors.

The Main International Modes

While in some areas of transport policy there have been quite conscious and deliberate attempts to devise co-ordinated, inter-modal approaches, this has seldom been the case with international transport. Even within the major trading communities (e.g. the EEC) despite lip-service being paid to co-ordinated policies, in essence policies have tended to be developed in the context of individual modes. In some instances this is understandable, substantial differences in costs mean that some modes so dominate specific markets that inter-modal co-ordination is simply unnecessary. There is little point in developing co-ordinated long-distance passenger travel across air, sea and rail when the former has such a clear absolute advantage in most circumstances. There are, of course, exceptions to this, of which the carriage of freight over land is perhaps the best example. Road and rail (and sometimes canals) often provide alternatives for transcontinental movements with even the possibility of hybrid modes (e.g. 'piggy-back' trains) being viable. Situations such as these mean that modes cannot be viewed in isolation from one another. Initially, however, we look at some of the areas where international policies are evolving with respect to single modes. In particular we look at international maritime and air transport policy and some emerging problems in international road freight transport. This is followed by a brief review of the policies of the

EEC where some degree of co-ordination is slowly emerging. This should be seen as a case study rather than as typical of the way all economic unions operate.

Shipping Conferences

International liner shipping (i.e. the provision of regular services over specified routes) has been subjected to cartelisation for over a century. While the first recorded 'conference' on the North Atlantic dates back to 1868, the agreement made in 1875 between seven UK shipping lines to regulate the London–Calcutta route represents the initiation of the type of system we are familiar with today. Essentially, conferences fall into two categories: the open conferences (found for example on the North Atlantic), where any lines may participate on the agreed terms, and the closed variety, where the route is restricted to those lines explicitly combining to form the conference. While the number of conferences, which are normally route specific and, in many instances, unidirectional, has fluctuated over time there are still well over 300 in existence.

Closed conferences, in particular, employ a variety of means to protect their market domination. Various forms of loyalty payments are offered to consignors to retain their custom while, if competition is threatening, the individual members may be given freedom to compete at market rates for carriage of those commodities which tramps are trying to attract away. Control over terminal facilities can also shield liners from unwanted market entry. Conferences may, however, enjoy natural benefits by having genuine economic advantages over tramp shipping. They offer regular services, for example, and take relatively small consignments which would not justify the hiring of a tramp ship. Availability of rate tables reduces search costs and empirical evidence points to a high degree of rate stability over time which facilitates easier long-term planning on the part of shippers (see Deakin and Seward, 1973). Operationally, market specialisation coupled with the overall size of most liner companies offers potential for greater efficiency in the supplying of shipping services.

Obviously, shipping conferences tend to point to the latter forms of benefit desirable from their collusion but they have, especially in recent years, met with recurring criticism. These generally relate to the profits they earn and the nature of the services they offer.

Comparative studies of the profits of liners and tramps over the

1950s and 1960s found there to be little difference, and where slightly higher returns were found to be enjoyed by the liners these could usually be attributed to efficiency factors. More recently, however, studies by Devanney, Livanos and Stewart (1975) and others have suggested that the comparison with tramp operations, even if taken over similar routes, is misleading. Given constraints imposed by conference rate setting, individual liners have tended to try and attract more tonnage to their vessels by providing higher quality services and, in particular, by utilising the most modern vessels. Hence, on average, conference costs are, despite potential scale effects, kept higher than they would be if the lines were free to compete with one another in a strict market efficiency sense. The appropriate analysis should, therefore, take as its basis the estimated outcome derived under conditions pertaining if there were free competition between the liners. This should then be employed as the counterfactual situation rather than the situation existing when there is competition between tramp shipping. Devanney *et al.*, conducting studies on efficiency basis of services between the USA and South American ports in the early 1970s, found conferences resulting in inefficiencies amounting to over \$20 per ton.

It is also being questioned whether the advantages of regular services, rate stability etc. really require cartel arrangements. If sufficient numbers of consignors are genuinely concerned about rate stability etc. then, even in a free market situation, forward markets in shipping services would soon develop or long-term contracts become more common. Equally, small consignments can be combined by forwarders to form full shipments which, in turn, permit all the economies usually associated with the employment of large ships to be reaped. Such arrangements would result in higher costs but these, it is being claimed, are likely to be lower than those associated with inefficiencies steming from cartel agreements. Further, they would be borne directly by consignors who explicitly require the services.

Just as there has been some shift in the way empirical studies assess conferences so recent years have seen something of a change in both national and international (mainly via the United Nations) policies towards their role. It is also worth noting that in some instances shippers have themselves combined into shipping councils to combat the monopoly strength of conferences via the exercising of countervailing powers (e.g. the European National

Shippers' Council rejected or substantially modified 18 of the 39 requests for general rate increases presented to them by non-US based conferences between 1975 and 1977). While these councils strictly represent a commercial rather than a policy response by governments, they often have a high level of official support. Given the knowledge and expertise of the consignors, many governments may well feel that fostering these commercially orientated groupings is likely to provide a more flexible counter-measure to the power of conferences than bureaucratic intervention on their part. Here our main concern, however, is with the more explicit policies which are developing with regard to the activities of shipping conferences.

One approach is to stimulate fleets to competition *against* the conference domination of a route by subsidising non-liner services. This is, in essence, the way that several COMECON countries are reacting, especially Russia and East Germany. (The People's Republic of China also has an expanding mercantile fleet.) They now account for something like 18 per cent of eastbound and 20 per cent of westbound traffic on the North Atlantic, a traditional preserve of liner activities, and enjoy a large share of the Mediterranean and Europe/East Africa routes. These market shares have been achieved by charging up to 40 per cent below conference rates. The COMECON ships enjoy both large capital subsidies and support from the regular business of state trading companies.

The impact of these COMECON tactics has been to force some liners to withdraw from established markets — their share of tonnage falling below critical break-even points — and to draw a gradual response from both national governments and economic unions (the EEC, for example, fearing an eventual domination by the COMECON fleet, are actively monitoring the situation, see Erdmenger, 1983). In the longer term it seems almost inevitable that more positive actions will follow as Western nations become concerned about their mercantile marine and the wider political implications of the COMECON expansion.

At the other extreme some nations, and most notably the USA, have acted to reduce the monopoly strength of shipping cartels by stimulating greater competition *within* the established regime. The USA introduced legislation as early as 1916 which exempts conferences from the anti-trust laws but required them to be open and to have their agreements approved by a maritime regulatory agency. More recently, in 1979, a series of investigations into rate-setting by liners operating from US ports resulted in legal action for failure of a number of liners to report schedules to the Federal Maritime Commission. The liners pleaded 'non-contest' and accepted a substantial fine to avoid the heavy legal costs of an extended, defended court action but, more importantly, it effectively meant the end of the Europe–North America Conference's ability to control either rates or capacity. The conference has in effect become little more than a lobby for shipping concerns and a forum for discussing problems confronting the liners.

On the rather grander scale major international initiatives have been set in motion. The United Nations, in particular, has been concerned with rectifying some of the imbalances caused by conferences to carriage between the less developing nations and the advanced, industrialised states. The specific situation perceived by UNCTAD was summarised in economic terms in a report appearing in 1969, viz:

> For many of the world's agricultural products, on which developing countries rely for much of their export earnings, supply elasticities are low in the short run . . . Although overall demand elasticities for most of these commodities are also low, the elasticity of demand facing the individual supplier or the whole group of suppliers in a single country is likely to be relatively high, unless that country is the only source of supply, and there is no ready substitute for the commodity . . . The supplier in these cases, therefore, normally bears the bulk of the transport costs (UNCTAD 1969).

Further, many Third World nations felt their own shipping activities were held back because their operators could not penetrate conference arrangements, which mean their potential earnings from shipping are severely limited. In 1978, for example, Third World countries only had 8.6 per cent of the world's deadweight tonnage but generated over 30 per cent of bulk cargoes and 90 per cent of tanker cargoes.

A response to this problem has been the drawing up of an UNCTAD Code of Conduct for liner conferences with the objective of establishing the so-called 40:40:20 principle (with 40 per cent of conference cargoes going by flags of the exporting country, 40 per cent of those of the importing country and 20 per cent by third flag ships — for more details see Neff, 1980). While Third World states rapidly accepted the scheme, the OECD and EEC nations received it less enthusiastically — they preferred monitoring conference activities and enhancing commercial freedom rather than pursuing policies of market allocations. After some internal discussion, the EEC agreed to allow members to accede to the Convention on the Code while retaining 'reservations' which would ensure commercial principles being retained for cross-trading within the OECD. Technically, conferences were basically given a group exemption from the cartel regulation of the community: this was a recognition, in fact, that market principles on occasions need to be, at least temporarily, abandoned to assist in easing the plight of the Third World.

Whether, in practice, less developed economies will be in a position to benefit materially from allocations of the kind contained in the UNCTAD code is much less clear. Evidence from 26 case studies conducted by Zerby (1979) found that only 9 of the countries examined had sufficient capacity in 1975 to handle 40 per cent of their exports and 16 sufficient capacity for 40 per cent of their imports. Expansion of capacity would seem likely, in the short term at least, to result in an excess of tonnage on the world market. Much of the cost of this may be pushed onto Third World fleets where, despite some of the trends we mentioned earlier, there are still serious imbalances between the *volume* of imports and exports which would result in a considerable number of empty sailings — in 1973, for example, Third World nations physically exported some 440 million tons of goods but only imported 242 million tons. In the longer term, therefore, there are likely to be factors which will contain the economic benefits enjoyed by Third World maritime nations but, despite this, it seems inevitable that political pressures (and especially the desire to 'show the flag') will be towards scarce resources being committed to expansions in their mercantile marine.

Whether the international co-ordination initiated by UNCTAD will continue to bloom in the future is going to be sensitive to a number of factors. By 1983, when the allocation system came into force, over 50 developed and COMECON countries had acceded to the code, representing about 23 per cent of the world's tonnage. While links between Europe and North and South America, Australia and New Zealand are still outside the system, important routes to Third World countries in the Far East, Africa and

Central America fall within the current ambit of the Code. More especially, some major traders are still outside the system (e.g. the USA and Japan) rejecting the control mechanisms inherent in any allocation procedure and particularly those which also involve cartelised suppliers. Some longer term arrangements offering compromise with the *laissez-faire* policies of these nations must be reached if the system is going to prove enduring. Equally, the general state of world trade is likely to play a crucial role — if excess capacity in shipping markets persists, and the issue of subsidised state-controlled fleets is not resolved, then it seems unlikely that a much broader geographical system will evolve during the period we are concerned with.

International Aviation

However it is measured, international air transport is the fastest growing sector in transport (see Table 5.3), being stimulated both by the increased emphasis being put on speed of service for freight carriage and by the enhanced desire for foreign travel which has accompanied higher disposable incomes in the West. On the supply side, major advances in technology, organisation and fuel efficiency have, over time, pushed down the real cost of air transport and expanded the scope of the services which can be offered. International co-operation is, by necessity, very considerable in this area both for flight safety reasons and to ensure optimal use is made of surface infrastructure. While each country may, under international law, prohibit use of its air space, a succession of conventions have established agreements about over-flights and technical landing rights. In other areas, however, usually relating to the actual scale of international operations, bilateral negotiations are periodically required.

   The so-called five 'freedoms of the air' date back to the Chicago Convention of 1944 (which established the International Civil Aviation Organisation) and the Bermuda Agreement of 1946. Bilateral agreements have since then traditionally been made between governments relating to the number of designated carriers which can fly to specific cities, the frequency of flights and the extent to which carriers can pick up passengers in one country for passage to points beyond in another country. The International Air Transport Association (IATA), an organisation of carriers, has typically set fares with governments having the right to object to particular fares filed.

Table 5.3: Trends in International Scheduled Air Services
(1970–80)

	1970	*1975*	*1979*	*1980*
Kilometres flown	2,530	2,900	3,530	3,610
Passengers carried	73,690	105,600	155,660	161,070
Passenger-km	159,360	263,590	430,700	457,060
Freight ton-km	6,300	11,460	18,690	19,990
Mail ton-km	1,440	1,160	1,350	1,440
Total ton-km	22,180	36,520	59,400	63,000

Note: Passengers carried — thousands; other statistics — millions.

Source: United Nations, *Statistical Yearbook*, 1981.

The system initiated at Chicago provided an organisational framework which met the needs of the limited international air transport market of the immediate post-war period but proved to have considerable limitations in terms of economic efficiency as traffic exploded in the 1960s and 1970s. Particular problems stemmed from:

(a) the unilateral restrictions on charter flights;
(b) the priority accorded to national airlines in terms of market shares irrespective of the efficiency of their services;
(c) the inflexibility of bilateral arrangements and notions of strict reciprocity when it came to the development of an air network.

Recent years have seen serious attempts to liberalise the system. The Bermuda II Agreement between the UK and USA was signed in 1977 and was initiated to liberalise fares policy. It has been followed by both similar arrangements — especially with regard to country of origin pricing — being reached between other European governments and the USA, and by a very dramatic drop in fare levels on North American routes (the latter being brought about mainly by US carriers withdrawing from the IATA fare-setting scheme to avoid national anti-trust laws). Numerous new routes were also initiated. This quite marked change in policy in the late 1970s brought about by public disquiet especially in the UK at the high fare levels then prevailing, coupled with the realisation that considerable latent demand existed for travel across the North Atlantic, has not been without its problems. Some fare regulation has had to be exercised by the US Civil

Aeronautics Board to smooth the transition to a more open market while there have been problems both within and between countries about the exact gateways which should be operated. At the detailed level the financial crash of Laker Airways and the problems associated with US antitrust laws, which led to the UK's Civil Aviation Authority refusing a number of cheap fares in 1984, have highlighted some of the remaining difficulties of the system.

Liberalisation on the North Atlantic, even accepting all of the accompanying teething troubles, has progressed at a pace far in excess of that of air transport within Europe. Here scheduled fares have remained much higher than on the North Atlantic routes (even allowing for the differing cost structures involved in operating overland and for the relatively short average flight distances) and constraints on gateway access are much more severe. One reason for this is that alternative European destinations are often seen as acceptable substitutes in the eyes of North American travellers; thus once fares fell on one route there was considerable pressure on others (this is essentially the mechanism which forced France to introduce lower trans-Atlantic fares). For shorter journeys within Europe, especially in relation to business travel, there are seldom alternatives to the preferred destination. Countries such as Germany have, therefore, found it possible to retain high fare levels and to favour the national carrier without fear of losing huge volumes of traffic.

Some liberalisation is still, however, gradually taking place with, for example, agreement between the Dutch and British governments being reached in 1984 that questions of capacity and fares on UK/Netherland routes would be left to the operating airlines. In these cases the bilateral managements between governments in effect often formalise changes which, *de facto*, occurred some time earlier. 'Bucket shops', for example, have been used by airlines to fill seats which would otherwise be empty at approved fares, and various methods of combining transport and accommodation have been offered to exploit loopholes in the regulations. In a sense, therefore, bilateral agreements are simply bending in response to prevailing market forces.

The EEC Commission is also pressing for some greater freedom within the Community. The final aim is not the complete commercial freedom of the type experienced since 1978 within the US domestic air transport market (national interests in state-owned carriers poses a serious political obstacle to the 'open skies' approach), but rather to achieve a proper balance between free competition and state intervention. While the shift of emphasis is unlikely to be rapid, it seems probable that the Community will increasingly become involved in developing measures that will gradually lead to greater flexibility in the traditional unilateral and bilateral organisation of air transport. The ultimate result is almost certainly going to be lower fares and greater variety in the types of service which are offered.

International Road Transport

As intra-industry trade is expanding between the developed nations and expansions are taking place in higher technology industries, so greater use is being made of road freight transport. Projections from organisations such as the EEC suggest that, under a wide variety of scenarios, both the relative share and aggregate demand for international road transport are going to continue to rise (combining Figures 5.1 and 5.2, for instance, provides substance for this line of forecasting).

Figure 5.2: Forecast Trends of Freight Transport in the EEC Based Upon Different Scenarios. Projected ranges expressed as percentages.

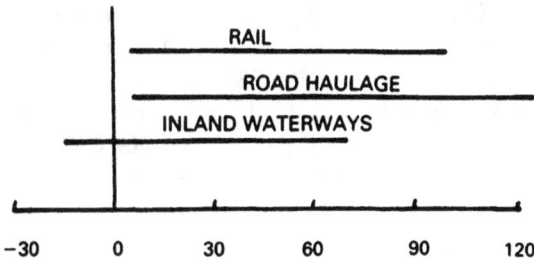

Source: Gwilliam and Allport, 1982.

This rise in the importance of road transport in certain international markets is posing a growing number of problems. The nature of road haulage and, in particular, the use it makes of national road networks have aroused concern from both outside the transport sector (e.g. from environmentalists) and within it (e.g. from competing modes and domestic road hauliers fearing competition from cabotage activities of non-nationals). In some cases there are institutional mechanisms (as, for example, in the case of the Common Market — see pp. 145–51 for some discussion of these) but often bilateral arrangements have to be made almost on an *ad hoc* basis.

An interesting case study of the types of problems which are arising can be found in the context of US/Canadian trucking. (Indeed, the situation in North America is even more interesting in the sense that each Canadian province is responsible for road haulage within its own territory — there are also minor technical differences in matters such as maximum vehicle weight.) Problems have arisen in this market because the liberalisation policy initiated in the USA under the Motor Carriers Act 1980 (essentially removing market entry restrictions based on criteria other than fitness to supply) contrasts sharply with the more restrictive policies of the Canadian provinces (especially Ontario and Quebec) most involved in cross-border transport. In effect, this opened the US market to Canadian hauliers while retaining the barriers to entry confronting US concerns trading into Canada.

Interestingly, and in part due to the relatively small share of overall US trucking moving across the border (amounting to about 2 per cent of the total revenue of the industry), this type of problem was not viewed as important when the deregulation was debated in America (see Button, 1985). Subsequent reaction on the part of the US road haulage industry, already hit in the early 1980s by national economic recession (tonnage carried falling by 15 per cent between 1979 and 1980 alone), was to press for reciprocal relaxation of entry controls in Ontario and Quebec and, failing this, to be given some protection from the Canadian undertakings moving across the border. The subsequent response of the authorities reveals some of the problems which may arise in international transport policy formulation with regard to ultra-flexible modes such as road haulage.

Complaints from the American industry that the Canadians were discriminating against it brought forth both a freeze on Canadian entry into the US market (enacted under clauses in the Bus Regulatory Reform Act of 1982) and an inquiry by the Interstate Commerce Commission into the issue. Negotiations were also set in train to try to reach some common policy with regard to cross-border traffic (although the lack of federal responsibility for such matters in Canada posed problems in this respect). The outcome was that the moratorium was subsequently removed. The ICC reported that the Canadian systems 'are applied in non-discriminatorily to American and Canadian applicants. Nevertheless, the disparity in the United States and Canadian entry standards puts American carriers at a disadvantage in competing for freight moving between the two countries.'

The negotiations brought agreement to establish consultancy mechanisms to avoid future conflicts. The short-term problems were partially solved for US undertakings by their buying up Canadian operators. Thus, while they achieved greater access to the Canadian markets, the actual level of competition within these markets was not substantially affected. Subsequently, however, provincial authorities in Canada have, despite approval at the federal level, refused efforts by the larger US carriers to expand their ownership of Canadian undertakings further.

Clearly, these types of conflict are likely to continue and increase in intensity as international trucking grows in importance throughout the world. Policies to handle the situation are likely to vary according to specific contexts but certain general principles would seem logical concomitants of any successful outcome. Firstly, from both EEC and US/Canadian experiences it is clear that agreement is unlikely to be reached on joint policies unless there is reliable statistical data available regarding the magnitude and detail of cross-border haulage flows. Information on this is often surprisingly scant and of doubtful accuracy. Secondly, many decisions relating to road-haulage policy are based purely on considerations relevant to the domestic market. This rather narrow view of the situation would seem to need widening if future policy conflicts are to be avoided. Finally, one must expect more permanent and carefully constituted forums to emerge in which international haulage policy may be discussed — previous *ad hoc* arrangements are unlikely to survive in the future.

The Common Market

International transport has certainly grown dramatically within Europe since 1965 (see Table 5.4) and we have already seen (Figure 5.1) that international transport is likely to represent the main area of transport growth within the EEC over the remainder of the century. The Economic Community has long recognised the importance of a Common Transport Policy (indeed it is one of

Table 5.4: Cross-Frontier Freight Traffic in Selected EC Member States. In 000s of tonnes

		1965		1980	Difference %
Railways					
Federal Republic of Germany	imp.	24,521		38,473	+ 56.9
	exp.	30,700		43,201	+ 40.7
France	imp.	18,454		22,361	+ 21.2
	exp.	30,480		25,920	− 15.0
Italy	imp.	20,575		29,689	+ 44.3
	exp.	10,086		11,314	+ 12.2
Netherlands	imp.	5,827		5,950	+ 2.1
	exp.	5,214		9,654	+ 85.2
Belgium	imp.	14,409		17,431	+ 21.0
	exp.	8,330		17,431	+109.3
Denmark	imp.	1,613		1,471	− 8.8
	exp.	775		754	− 2.8
Road transport					
Federal Republic of Germany	imp.	15,723		56,566	+259.8
	exp.	8,997		52,601	+484.7
France	imp.	7,540		35,960	+376.9
	exp.	9,860		37,100	+276.3
Italy	imp.	2,607	1975:	10,011	+284.0
	exp.	3,872	1975:	11,299	+191.8
Netherlands	imp.	5,829		23,528	+303.6
	exp.	6,863		23,393	+240.9
Belgium	imp.	7,457		35,441	+375.3
	exp.	9,093		40,184	+341.9
Denmark	imp.	1,025		4,257	+315.3
	exp.	2,360		5,340	+126.3
Inland waterway transport (only in millions tkm, international traffic)					
Federal Republic of Germany		18,700		27,949	+ 49.5
France		2,752		3,808	+ 38.4
Netherlands		12,313		20,489	+ 66.4
Belgium		3,098		3,743	+ 20.8

Source: Statistical Office of the European Communities (ed.) (1982), *Statistical Yearbook for Transport, Telecommunications, Tourism 1980* (Luxembourg), pp. 23, 41, 73; own calculations.

only four areas where an explicit common policy is specified in the founding Treaty of Rome) in facilitating the efficient working of the market. The difficulty has been in deriving a policy which is acceptable to all member states. The essential difference between the Anglo-Saxon approach favoured, for instance, by the UK with its emphasis on economic efficiency within the transport system contrasts with the socially orientated philosophy of countries such as France where strict efficiency within transport is viewed as secondary to the wider role transport may play in economic planning. On the one hand, the Anglo-Saxon approach, therefore, seeks for greater market freedom with regulations designed to minimise imperfections in a competitive environment. On the other hand, the traditions of the *Code Napoleon* require regulation as the basic policy with market forces only allowed a free rein when they assist in the attainment of wider goals.

The differing philosophical positions of member states have been accompanied by a variety of other factors which make co-ordination difficult (see Button, 1984). If we look at Table 5.5, for instance, it is clear that different member states rely upon different modes to transport their freight. There is a tendency, therefore, to advocate policies which would help the international expansion of that mode in which they have the greatest strength (hence the Dutch and British have pushed policies favouring greater freedom for international road haulage movement in the Community). The situation is further complicated by the fact that, from a transport perspective, the EEC does not form a natural market. Greece, is, for instance, geographically separate and obvious trade routes between other members pass through third party states (Switzerland and Austria). One implication of this is that EEC transport policy must be set in the context of arrangements for the larger European transport system (e.g. those devised by the European Conference of Ministers of Transport and the Economic Commission for Europe of the United Nations).

Given these difficulties it is not surprising that a genuinely Common Policy still eludes the Community after over twenty-five years of trying. The attitude of the Commission which attempts to define policy for ultimate ratification by the Council of Ministers has changed quite markedly over this time (see Erdmenger, 1983). Initially, grandiose blueprints were drawn up offering an overall set of measures designed to harmonise the transport systems of member nations as well as to facilitate improved transport supply

Table 5.5: Percentage of Tonne-Kilometres of Freight Carried by Mode in Selected EEC Countries in 1978

	Road	Rail	Water ways	Pipeline
France	44	34	6	16
West Germany	43	27	24	6
Italy	76	15	0.5	9
United Kingdom	77	15	0.1	8

between members — e.g. the Schaus Memorandum of 1961 and the Action Programme of the following year. While some of the individual measures being advocated, albeit often in severely modified forms, were gradually initiated the overall package was too ambitious. Harmonisation of internal transport on the scale suggested was, for example, unacceptable to many members while others, most notably Holland, objected to policies involving rate regulation and vehicle weight limit standardisation.

Progress had been so slow by the time of the enlargement in 1973 (when the UK, Denmark and Eire joined) that the Commission Member for Transport said in an address to the European Parliament that

> Without wishing to detract from progress already achieved, particularly when it is viewed in relation to existing difficulties, we must nevertheless frankly and objectively admit that few of the aims of the Common Transport Policy have been achieved.

The 'New Impetus' that followed was coloured by the more commercial orientation of the additional member states plus a more general appreciation that the large scale, blueprint approach of the Commission in the past needed revision. Organisation changes within the Commission itself partly reflected the modified attitudes with administrative divisions generally being based on policy issues rather than being exclusively mode specific. The key change in emphasis within this new framework was the relaxation of concern over state intervention in transport and a movement towards aligning such intervention at the Community level, especially with regard to infrastructure and market organisation, in such a way that Community interests are best served. Concern with rate setting and entry controls gave way to greater emphasis on co-ordination in infrastructure provision and pricing. At the operational level the Community was seen increasingly as an agent for intervention at times of crisis. Continued market monitoring would indicate when the smooth running of the market was being disrupted and action required. Further, transport was linked to other key areas of Community interest (e.g. energy, the environment, regional policy etc.) and policy was to be developed in a much broader integrated framework.

Progress under the New Impetus was not exactly rapid and, indeed, by the late 1970s commentators were pointing to the failure of the system (Gwilliam (1979), for example, typified the position of many when he said 'Progress in the creation of a common transport policy in the EEC has always been slow. Since the accession of the new member states in 1973 it has come to a standstill.') Part of the problem lay in the fact that member states perceive very few direct benefits to themselves from pursuing the types of policy contained in the New Impetus — there may be Community gains but these are difficult actually to quantify and even more difficult to allocate to individual states.

In fact, the approach of the Commission has not been static since 1973 but rather has been evolutionary with an increasing emphasis being placed upon pragmatism. Indeed, actions by the European Parliament to take the Council of Ministers before the European Court of Justice in the early 1980s, if more rapid progress were not made towards devising a more thorough Common Transport Policy, resulted in a succession of measures being taken regarding such things as considerably enlarging the quota of multilateral road haulage permits and simplifying documentation needed for border crossings.

For substantive progress to be made in the future the Commission (and the Council) will need to modify their approach. Progress in other sectors, such as agriculture (i.e. the Common Agricultural Policy), has been possible because direct public intervention with the market mechanism to achieve specific, narrow goals was involved. With transport it is a question of structural change, which is not in itself easy but is even more problematic if, as in this case, key participants (such as France and West Germany) do not perceive serious distortions in the existing structure. Ways forward may be found by concentrating on partial or short-term experimental approaches (an idea, for instance, favoured by Erdmenger, 1983) which cover specific, limited geographical areas of the community or which operate only for a specified period. Any benefits derived from such initiatives are likely to be more apparent than hypothesised gains shown up in desk-top studies and could thus lead to their wider adoption. In a similar, but slightly different, vein Gwilliam (1979) suggests that the Commission should concern itself with identifying areas where problems exist and where members perceive the need for common rules to be devised. (COMECON trade, energy conservation and research matters may well fall within this.) There should be no attempt, however, to *enforce* a common structure in such situations. The Commission should also steer and help the Council and members in cases where common rules already exist or agreed principles have been arrived at. In the latter case it should examine the extent to which individual countries' transport policies conflict with the common principles and, where serious deviations exist, the Commission should initiate negotiations to resolve conflicts (vehicle weights, through traffic and questions of border crossing may fall under this heading).

Both of these suggestions involve a much more limited idea of a Common Transport Policy. They also have their weaknesses. Erdmenger's approach ignores the potential distortions which may arise if a policy is only applied over a limited area, while transport suppliers are put into a world of increased uncertainty if they are unclear as to the degree of permanency of a policy measure. Equally, with Gwilliam's approach, there are few areas where all members perceive problems to exist and where notions of common rules are sufficiently close to bring about rapid agreements. The idea of focusing on the formulation of general principles has attractions but arriving at such principles is unlikely to be easy unless they are of such a vague form as to be worthless in a negotiating context.

These types of approach, combined with the continuation of the pragmatic attitude of the Commission, are, however, likely to be the characteristics of EEC transport policy over the next decade or so. The expansion of the Community to embrace Spain and Portugal is likely to add to the problems of making any serious progress in the traditional fields of land transport.

In addition to the periodic geographical expansions which the Commission has to contend with, there are also new areas of transport which have recently come to the fore. Air transport policy and maritime transport have become more central in recent years (see also the previous section). Integration of these modes into the overall transport policy is taking place but their importance in much wider transport markets will inevitably make slow progress.

Some Conclusions

What seems to be emerging in the field of international transport is the *ad hoc* and partial nature of the measures which are being taken. Actions tend to be taken more in response to crisis situations rather than as part of a co-ordinated, proactive programme. There seems to be little evidence that this situation is likely to change dramatically in the foreseeable future. Certainly a number of trading communities have attempted to develop something along the lines of a unified policy for transport between members, but the outcomes seem little more comprehensive than the agreements (both multilateral and bilateral) that have developed between nations which act outside such groupings.

Part of the difficulty, as we have seen, stems from the inevitable tendency for national governments to protect their own manufacturing industries and to maintain their own transport undertakings. It is only when there seems a mutual benefit to be derived by all parties that countries come together to tackle problems on the international front. Even then, however, agreements are often difficult to reach. While part of the problem often lies in disagreement over the nature of common actions, governments being concerned with ensuring that their share of benefits are maximised, it also stems from fundamental differences in national attitudes to transport policy. Compromises may be found which satisfy the minimum demands of each country with respect to the detailed nature of co-ordination but where there are conflicts in philosophy to resolve, the situation often proves intractable. Looking to the future there seems no reason to suppose that the basic conflicts between those states which take a purely commercial attitude towards transport and those adopting a wider, socio-economic perspective will be more easily reconciled than in the past. Indeed, given the probable growth in trade coupled with changes in its nature and composition there are likely to be even more problems seeking resolution.

As a final thought and comment, it is perhaps worth remembering that while there are problems in finding satisfactory forms of transport co-ordination at the international level, many countries have trouble devising internal transport policies in circumstances where there are important regional variations or where the administrative system is of a federal kind. Given the problems confronting them and the multiplicity of different 'horse-trading' combinations to be performed, it is perhaps surprising that international politicians have achieved what they have.

References

Button, K. J. (1984), *Road Haulage Licensing and EC Transport Policy* (Gower, Aldershot).

Button, K. J. (1985), 'Transport Deregulation and Cross-Border Trucking in North America, *Transportation Planning and Technology*, vol. 9, pp. 285–99.

Deakin, B. M. and T. Seward (1973), *Shipping Conferences: A Study of Their Origins, Development and Economic Practice* (Cambridge University Press, Cambridge).

Devanney, J. W., V. M. Livanos and R. J. Stewart (1975), 'Conference Rate Making and the West Coast of South America', *Journal of Transport Economics and Policy*, vol. 9, pp. 154–77.

Erdmenger, J. (1983), *The European Community Transport Policy: Towards a Common Transport Policy* (Gower, Aldershot).

Gwilliam, K. M. (1979), 'Institutions and Objectives in Transport Policy', *Journal of Transport Economics and Policy*, vol. 13, pp. 11–27.

Gwilliam, K. M. and R. J. Allport (1982), 'A Medium Term Transport Research Strategy for the EEC: Part 2 — Research Priorities and Organisation', *Transport Reviews*, vol. 2, pp. 349–72.

Helleiner, G. K. (1973), 'Manufactured Exports from Less Developed Countries and Multinational Firms', *Economic Journal*, vol. 83, pp. 21–47.

Neff, S. C. (1980), 'The UN Code of Conduct for Liner Conferences', *Journal of World Trade Law*, vol. 14, pp. 398–423.

UNCTAD (1969), *Level and Structure of Freight Rates, Conference Practices and the Adequacy of Shipping Services* (United Nations, New York).

Zerby, J. A. (1979), 'On the Practicability of the UNCTAD 40–40–20 Code of Liner Conferences', *Maritime Policy and Management*, vol. 6, pp. 241–51.

6 TRANSPORT AND DEVELOPMENT

Economic Development

Development has many dimensions and can take a variety of forms — social, economic, political, cultural etc. In this chapter, however, we focus almost exclusively on economic development although, as we see below, this in itself is often not a straightforward concept. This is not to say that transport policy is not important in other respects and certainly the improved access provided by enhanced transport facilities has resulted in a significant spread of Western ideas, medicine and culture throughout the world. Whether this should always be seen as 'development' is, of course, a point of debate but certainly there has been some form of impact. It is also important to appreciate that, in some cases, these transport-induced changes in culture have produced significant indirect economic changes. Ideas of the market economy, contact with Western consumer goods, improved medical treatment etc. have all had an impact on the supply of labour in Third World countries and have opened up new markets for industry. Generally, however, these longer term developments are difficult to measure and even more difficult to relate to specific transport facilities. The result is that transport policies have normally been viewed in terms of their immediate impact on an area's GNP or GDP rather than their longer term implications for the local economic system.

Within a country the notion of economic development may have distributional elements as well as being exclusively seen in terms of National Income Accounts. These elements may involve ensuring that targeted groups within the country enjoy a certain share of any rise in GDP but, from our perspective, more importantly they may embrace considerations of the geographical spread of the increased income. Many would argue that a rapid rise in the economic well-being in some regions while others stagnate or decline (a so-called dual-economy effect) does not represent real national development at all. In this sense, the specific economic development of lagging or depressed areas may be seen as a key policy target even if this means that income growth elsewhere is held back. Such cases may involve the use of transport policies to ensure that economic growth spreads to these less obviously dynamic regions or cities so that more equitable economic development may be achieved. This is the basis of nationally organised regional economic policies — see below, pp. 159–68. In a wider sense, it is also the basis of internationally organised regional economic policies (such as those of the EEC regional policy) which transcend national boundaries.

The purpose of this chapter is to examine the role that transport policy may play in aiding and directing economic development. The notion of direction is an important one, as we have seen above, and in many cases is, in fact, inseparable from the more conventional concern with growth *per se*. For political and economic reasons it is often necessary to ensure some degree of equity in the geographical spread of development to ensure stability in the country and to utilise fully the nation's resources. In assessing policy, however, it is convenient to treat the two issues independently although potential interactions should be borne in the back of the mind.

Our approach is initially to examine the links between transport provision and economic development. This is, the reader should be warned, a topic which has been the subject of considerable debate and is certainly still far from being fully understood. (The lack of understanding has, of course, not prevented quite doctrinaire positions being taken by policy-makers at various times.) Some general discussion of the approaches pursued to date is then offered. While much of this concentrates on the role of transport in Third World development strategies there is still a substantial coverage of policies favoured in more industrialised countries. Future trends in transport policy with respect to economic development are not, and indeed cannot, be viewed in isolation both from the overall objectives of future development policies and from the prevailing, underlying philosophies of economic planning as a whole. Transport must be seen as only one component of a much wider set of instruments which are employed by policy-makers in their quest for economic advancement.

Transport and Economic Development

Transport has long been linked with the process of economic development, be it at the national or local level. It is not surprising, therefore, that transport policies have often, at least partly, been viewed as having development objectives. Certainly most development plans of Third World countries embrace proposals to direct substantial investment resources to the transport sector (see Table 6.1) and earmarked sums for transport schemes constitute a large proportion of the resources made available by international agencies such as the World Bank, the Inter-American Development Bank and Asian Development Bank. (E.g. the World Bank allocated US$ 1,614.2 million in 1982 to transport projects representing 12.4 per cent of its total budget). In fact, however, the actual relationships between transport provision and economic development are still little understood. While it seems logical that one should not entirely separate transport issues from the location of economic activity or divorce them from spatial policy problems, the detailed connections are often difficult to specify and even more problematic to quantify.

Table 6.1: Transport Investment as a Percentage of Total Planned Public Sector Investment in Selected Third World Countries

Country	Plan	% Transport
Mauritius	1971–5	17.5
Tanzania	1969–74	28.9
Kenya	1974–8	40.6
Sudan	1970–5	14.8
Botswana	1973–8	26.2
Zambia	1972–6	29.7
Uganda	1971–5	28.2
Swaziland	1973–7	17.0
Malawi	1971–80	34.8
Senegal	1973–7	17.5
Sierra Leone	1973–8	21.3
Nigeria	1975–80	27.5
Philippines	1971–4	57.5
Thailand	1972–6	19.5
Malaysia	1971–5	18.4
India	1974–8	19.2

Source: S. Thomas (1977), 'Road Investment and Pricing in Developing Countries', *Oxford Bulletin of Economics and Statistics* vol. 39, pp. 203–17.

Thinking on these matters has tended to change quite dramatically from time-to-time. The traditional view, which stemmed mainly from the work of economic historians and held sway until the post-World War Two period, was that by opening new markets and easing access to raw materials, the costs of production could be lowered as economies could be reaped from both larger scales of production and the greater abundance of inputs. Transport was seen as a driving force behind economic development and, for example, transport innovations such as canals and metalled roads were perceived to be major pillars upon which the Industrial Revolution was built. Some have taken an even stronger position and seen transport improvements as the key factor stimulating economic growth. Perhaps the classic view along these lines is that of Rostow (1960) with respect to rail expansion, viz., 'The introduction of railroads has historically been the most powerful single indicator to [economic] take-offs. It was decisive in the United States, France, Germany, Canada and Russia.'

A logical extension of the traditional view is that economic development, especially in Third World countries, can only be achieved by considerably improving their transport infrastructure. In the 1920s, for instance, Lord Lugard was arguing that 'the material development of Africa may be summed up in one word — transport'. Even in Western countries the view still persisted that improved transport would stimulate growth and a considerable amount of the investment in rail and inter-urban road infrastructure in the post-war period has explicitly been justified on development criteria.

The idea that transport can act as a major catalyst for subsequent economic expansion began to be questioned in the 1960s — in part, one suspects, because of a rising general scepticism over the justifications being advanced for the vast sums then being channelled into prestige transport investment projects. Economic historians began to argue that the available evidence suggested that major transport innovations, such as the canals and railways, had actually lagged behind spurts in economic growth rather than preceded them. Counter-factual studies, of which the work of Fogel (1964) on US railways is perhaps the best known, began to appear employing econometric procedures to generate pictures of the probable growth path of economies if transport changes had not occurred. The 'new economic history', while open to several lines of criticism, has generally shown transport to be much less important in the development process than was implied in the traditional model. Where it has revealed transport to be a major influence it is only in the context of very specific circumstances and in conjunction with other changes.

These latter studies imply that transport is often reactive rather than proactive and that it may be more reasonably seen as a factor permitting growth rather than actively generating it. Inadequate transport is viewed as a bottleneck to the full exploitation of the economic potential (especially in the agricultural sector) of the country or region. It is only *one* potential bottleneck, however, to socio-economic development which should be examined on a more-or-less case-by-case basis. If one accepts this view then very high levels of investment in transport infrastructure are often likely to prove wasteful if the other prerequisites for growth are not also achieved.

There is a further consideration. Even if transport provision is capable of stimulating economic growth in certain social environments it may have no impact in others. People vary in their perceptions and some simply do not perceive the opportunities which confront them. Others, for a variety of reasons, may recognise the potential but decide not to exploit it. In some ways these factors may be viewed as similar to the non-transport bottlenecks outlined above but they are of a much more fundamental nature and require an altogether different approach from policy-makers.

While these latter positions have gained ground in recent years there is still, nevertheless, a proclivity for many less developed countries to see improved transport as a panacea to their development problems. In part this stems from a tendency to view physical structures such as airports, highways and dock installations as actual manifestations of advanced development in their own right. Prestige is no small consideration. Additional to this and reinforcing it, planning agencies and planners themselves who are responsible for the allocation of public investment funds have had a tendency, for several reasons, to favour policies involving projects with a long economic life and for which the overall costs and benefits are difficult to estimate. This gives them enhanced prestige and provides a certain mystique to their decision-making but it also means that it is difficult actually to prove that mistakes have been made. The introduction of outside expertise into these situations seldom does anything to reduce the problem. Most consultants have Western backgrounds where, because development is more advanced, heavily capital-intensive programmes are usually optimal and where the impact of mistakes, in the face of relatively abundant resources, is much less damaging. This latter effect is compounded by the fact that much aid is tied by the donor country which, in turn, is confronted by its own industrial sector wishing to find a market for its state-of-the-art technology.

International bodies, such as the World Bank, have in recent years tended to tread a rather careful path between these two schools of thought and have attempted to adopt a more pragmatic approach. Thus while the World Bank argues that 'Transportation is a necessary concomitant of the exchange economy and is indispensable to economic growth' it goes on to say, 'The safest strategy of investment institutions is to wait until increases in production or capital productive schemes signal clearly the infrastructure requirements' (World Bank, 1972). Despite this selectivity the World Bank still puts considerable sums into the transport sector although the emphasis has now switched from an early emphasis on high-profile, prestige projects to less spectacular rural road schemes designed to improve accessibility.

While one aspect of this change can be viewed in strict economic terms it is also true that, irrespective of the impact on national income, improved road transport can have important effects on the lifestyle of those brought into closer contact with developed sectors of the economy. This point was clearly made some years ago by Wilson (1966), viz.,

> Investment options might usefully be analysed in terms not only of their direct economic pay-off but also in terms of their influence on attitude . . . The educational and other spillover effects of road transportation appear to be greater than those of other modes of transport. This is especially significant at low levels of development.

Economists have also increasingly moved away from the idea that one should treat transport and development in aggregate terms and have separated the impacts of transport policies on specific industrial developments from that on economic growth *per se*. Transport is seen to influence the success or otherwise of firms at the margin. Whether this results in a dramatic impact on the economy as a whole then depends upon how many firms or sectors actually fall within this margin. A major transport investment may, for example, have a dramatic impact on an industry which is isolated from its markets but otherwise enjoys high levels of efficiency. If other industries in the region or country, however, are restrained by poor efficiency records or are employing outdated technologies then even massive transport investments are not going to produce significant economic expansion. In the jargon of economics, transport policies may have important effects at the margin but need not significantly improve the *average* growth rate in the country. Many economists are, therefore, sceptical about the generalisations which are often made regarding the automatic desirability of transport improvement. In many ways these ideas tend to accord with the more cautious approach of the World Bank and with the emphasis on a case study philosophy, when one takes particular cognisance of the alternative ways the resources could be employed.

There are also wider grounds still for wishing to improve transport in the Third World. Many people quite simply think that it is wrong that countries with 75 per cent of the world's population account for only 13 per cent of railway freight, enjoy only 10 per cent of its paved roads and only 9 per cent of its motor vehicles. Simple justice would seem to indicate that something needs to be done when a dozen countries accounting for half the world's population have only one-fifteenth of its motorised transport equipment.

Transport and the Spatial Distribution of Development

In addition to being viewed as instrumental in the stimulation of economic growth *per se*, transport has also, as we have emphasised in the previous section, often been attributed with influencing the geographical distribution of economic activities. It is seen as a prime mover in the location as well as in stimulating the aggregate level of economic performance. (However, it is worth noting that some economic theories of the development process suggest that an even balance — across both space and industrial sectors — is the most efficient way to proceed. If so, then the distributional issue is very closely entwined with the question of national economic growth — the debate over the validity of this view is still, however, going on.)

Producers, it is claimed, with mobile plant would prefer to locate, other things being equal, in areas which offer the best access to markets and the easiest route to inputs. Policies have, therefore, regularly been presented which are intended to redirect economic activities (usually, on distributional grounds, to areas of high unemployment but occasionally for reasons of political power) to favoured locations by means of enhancing transport links with the rest of the economy. This type of thinking, for example, typified post-war transport policies in the post-1945 period with it providing a key argument for the Inter-State Highway System in the USA, the north–south autostrada links in Italy and for the construction of the UK motorway network.

A look at the history of trunk road and motorway building in the UK is particularly revealing in this respect. The Barnes' Plan of 1946, for example, cites 'improvements to assist development areas in particular and industrial development generally' as a specific objective of the post-war inter-urban road building programme. The same theme re-emerged in the subsequent programmes of Conservative and Labour administrations, although the late 1950s witnessed debate about exactly which regions should 'benefit' from improved road links (e.g. see Starkie, 1982). In the 1960s we find the Labour government pursuing objectives of providing (in 1969) 'vital access between regions and between major cities' (Starkie, 1982) and stating that 'Particular attention has been paid to the support that new and improved roads give to the major measures that the Government are taking to ensure a better balance in economic development' (in 1970) (Starkie, 1982). We find the same view extending into the 1970s with the Conservative government of 1970–4 intending 'to link the more remote and less prosperous regions with [the] new national network' and, later, 'we need to provide new and improved roads . . . to aid our economy, including regional development' (Starkie, 1982).

This tendency to attempt to assist depressed regions via transport investments is also revealed if one looks at past public expenditure patterns in the UK (see Table 6.2). When national accounts are broken down on a regional basis it is seen that in the UK the traditionally depressed areas (Wales, the north, Scotland, Northern Ireland, Yorkshire/Humberside and the north-west), in general, get more than the national average of transport expenditures.

The exact strength of the arguments for transport investments, despite their immediate intuitive appeal, are not, however, accepted by everyone. Certainly there are likely to be sectors and industries which are very strongly influenced in their choice of location by the prevailing quality of transport infrastructure but equally, it has been argued, many are much less sensitive. The question should, in the view of these dissenters, be asked whether investing in the transport system is the most effective use of resources given the locational objectives being pursued. The gen-

Table 6.2: Regionally Relevant Public Expenditure 1977/78. Per capita index UK = 100

	North	Yorkshire/ Humberside	East Midlands	East Anglia	South East	South West	West Midlands	North West	Wales	Scotland	Northern Ireland
Agriculture, forestry and fisheries	107.3	75.6	75.6	115.0	61.1	100.0	69.9	60.6	142.5	229.5	382.4
Trade, industry and employment	219.1	87.2	67.1	30.0	46.5	55.1	78.6	118.5	192.2	179.4	334.2
Roads and transport	116.8	99.4	65.5	94.9	107.7	70.0	76.2	88.9	124.7	136.2	120.6
Housing	109.0	76.1	84.3	77.7	125.6	55.3	94.6	89.1	73.7	117.4	118.5
Other environmental services	125.5	86.6	72.2	88.1	103.1	69.9	84.5	97.7	111.3	156.9	112.8
Law, order and protective services	87.2	88.1	79.8	75.0	110.2	86.9	80.7	83.5	82.1	89.8	329.5
Education, libraries, science and arts	96.6	97.6	90.6	90.2	105.2	82.4	87.3	97.7	103.0	119.3	122.2
Health and personal social services	94.5	90.6	86.5	86.0	107.7	93.2	85.0	98.9	98.6	118.7	119.2
Social Security	109.7	102.6	92.0	87.1	92.9	103.3	92.8	109.6	116.1	105.9	109.5

Source: J. Short (1981), *Public Expenditure and Taxation in the UK Regions* (Gower, Aldershot).

eral view of this group of eclectics is summarised by Gwilliam (1979):

> at best, transport infrastructure investment as an instrument of regional policy could only be effective in particular circumstances (where the less active region was a relatively efficient producer of tradable goods) or as part of a package (the rest of which must have the effect of reducing local production costs in the target regions below those in its potential trading partners' economies).

A similar set of conclusions is drawn by Sharp (1980) in his study of transport policy in Britain. Essentially he contends that improved transport is unlikely to cause significant economic activity to take place in spatially-disadvantaged areas but that it may be a necessary concomitant to other measures if these are to be successful. In particular, if state intervention is to result in new industrial growth in a peripheral area, 'then low-cost inter-regional transport facilities must be provided'.

The influence of these views is now beginning to permeate official thinking. We find this, for instance, in a number of official enquiries. The Leitch Committee (Department of Transport, 1978) looking into the effects of trunk road investment, for instance concluded:

> that trunk road construction does not yield significant economic development gains over and above the direct benefits to road users . . . The only exception is likely to be where new construction links previously separate regional or national networks giving very large reductions in generalised costs.

Equally, the Armitage Committee (Department of Transport, 1980) found that:

> When industry and commerce make decisions about the location of factories or their system of distribution, it is often less important to reduce transport costs than to reduce other costs such as those of stockholding or to take advantage of the grants for setting up factories in assisted areas.

Recalculations of Table 6.2 may, if indeed these recommendations are made effective, look somewhat different in the future.

The difficulty with accepting this position, from a policy point of view, is that it would seem to imply that transport investment should be concentrated in the already prosperous areas. Congestion in these areas is a significant cost and transport investments would reduce this. Two immediate points should, however, be made in this respect. Firstly, the argument that funds be directed to parts of the system currently experiencing heavy congestion ignores the actual controls over the system which are being exercised. If one introduces environmental costs and congestion costs into the accounts of manufacturing firms (e.g. by making them pay directly for the roads they use or by making them compensate sufferers for environmental damage they inflict) then relocation to peripheral areas may become more attractive and transport investment away from the congestion regions become (on strict economic grounds) efficient. Secondly, the argument rests exclusively on price-efficiency criteria and ignores questions of equity — social fairness may itself be deemed sufficient justification for investing in depressed areas.

The question of the effectiveness of transport policy as a spatial economic policy tool is also one requiring a little more consideration. Empirical studies of this question may help but here there are serious problems — in many ways one would hardly expect the effects of improved transport provision to be easily calculated. Most of the studies which have been completed are, mainly because of data problems, of a very aggregate kind. (Essentially, one can only disaggregate below the level of broad regions and standard, official industrial classifications by accepting a substantial reduction in statistical reliability.) The level of aggregation may, however, be of major importance. For example crude calculations at the national level (see e.g. Table 6.3) suggest that transport costs only account for about 9 per cent of total production costs but further breakdowns indicate a range of about 20 per cent in the importance of transport costs to manufacturing industry by region. Transport may, therefore, be important at a spatially and industrially disaggregate level which is lost when looking at broad statistical groupings. Disaggregation may take a variety of forms and, from the point of view of policy formulation, since each is likely (for the reasons we have cited above) to have some limitations it is helpful to look at some of the ways this has been done. In particular, it is important to ask if there are specific industries which are sensitive to transport considerations when location choices are made.

Table 6.3: Transport Costs for a Selection of British Industries in 1963 as a Percentage of Net Output

	Total census transport cost (% of the value of net output)
	Low
Engineers' small tools	1.75
Office machinery	1.59
Ordnance and small arms	1.17
Scientific instruments	1.65
Watches and clocks	1.08
Telegraph and telephone apparatus	1.44
Radio and electronic apparatus	1.62
Aircraft manufacture	1.06
Locomotives	0.86
Railway carriages	1.49
Cutlery	1.21
Jewellery	1.69
Lace	1.97
Men's tailored outerwear	1.94
	High
Coal mining	25.24
Stone and slate quarrying	39.78
Chalk, clay, sand and gravel extraction	29.78
Metalliferous mining and quarrying	18.63
Confectionery	24.71
Milk products	27.54
Sugar	24.24
Animal and poultry foods	17.04
Soft drinks, British wines, cider and perry	23.76
Coke ovens and manufactured fuel	19.95
Fertilisers and chemicals for pest control	17.40
Bricks, fireclay and refractory goods	22.27
Miscellaneous building materials	22.06

Source: Edwards, 1970.

At first sight it would seem that those industries whose location decisions are most sensitive to transport provision are those which, given that they are not constrained by physical factors (e.g. some extractive industries), require a substantial transport input into their production activities. If we examine the UK work carried out by Edwards (1970) and US work by Anderson (1983) on the importance of transport costs in overall costs, some of which is summarised in Tables 6.3 and 6.4, we find that industries such as cutlery, office machinery, menswear, radio equipment etc. are likely to be less influenced in their choice of location than, say, soft drinks, milk products, confectionery, animal foods etc. (Indeed, drawing on US experience, it has been suggested that Gary, Indiana was specifically favoured by US Steel because it minimised transport costs — in Britain the transport cost component of steel is about 7 per cent.) The important fact stemming from this information, especially from the point of view of spatial location policy, is the proportion of industry which falls into the transport cost sensitive band. Superficially this seems to be relatively small, at least in the UK. Gudgin (1978), for instance, estimates that in 1968, 'almost three quarters of British industry incurs total transport costs at levels of less than 3 % of the value of gross output. In 95 per cent of industry, by value of production, the transport costs are less than 5 % of total costs.'

Table 6.4: Transport Costs for a Selection of US Industries in 1982 as a Percentage of Product Prices

		Inbound and outbound transport costs (% of product prices)
High	Stone, clay, and glass products	27
	Petroleum products	24
	Lumber and wood products	18
	Chemicals	14
	Food and kindred products	13
	Furniture and fixtures	12
Medium	Paper and allied products	11
	Primary metals industries	9
	Textile mill products	8
	Fabricated metal products	8
	Miscellaneous manufacturing	8
	Transportation equipment	8
	Rubber and plastics products	7
Low	Tobacco manufactures	5
	Machinery excluding electrical instruments	5
	Instruments	4
	Apparel and other textiles	4
	Printing and publishing	4
	Electrical and electronic machinery	4
	Leather and leather products	3

Source: Anderson, 1983.

These facts go a long way to explaining some of the findings of statistical studies carried out on specific impacts of transport changes. The study of the M62 (Lancashire–Yorkshire) motorway, for example, suggests that the distributional effect of the investment was not substantial. Only some 2,900 jobs changed location per annum as a result of it out of a total labour force in the area of 3.4 million (see Judge, 1983).

Several interview studies conducted in various European countries offer considerable support for these views. Local authorities in Italy found that only 10 per cent of business located in the south had, when asked, pointed to the improved motorway system as a dominant factor, while in France similar work produced no tangible links between motorway provision and new industrial developments. A study by Fischer (1971) superficially seems to offer counter evidence in that 41.5 per cent of firms relocating in West Germany over a seven-year period cited road provision as an influence on their decision but closer inspection reveals it to be only fourth in the list of overall priorities.

While these data and interviews suggest that transport policy may have a very small effect on industrial dispersion one or two caveats need to be mentioned. Firstly, the objective of regional policy is often to assist in job creation. Measurements relating transport costs to production costs may, in these circumstances, not be the appropriate set of calculations to examine — labour-intensive industries may be more sensitive to transport costs in their location decisions and this needs more research. Secondly, the use of production cost ratios in much of the statistical work (especially expressed in terms of average rather than marginal costs) ignores the other side of the equation, namely revenue. Even if financial criteria are used to examine the strength of transport policy in regional strategies it seems more logical to look at the impact of public transport investments on net profits. Chisholm (1971), for example, suggests that transport costs may represent 25 per cent of the profits of manufacturing industry in the UK. Thirdly, financial costs are not necessarily the key parameters considered by industrialists anyway when reviewing transport inputs — speed, reliability and security have long been of major importance. A succession of studies in the past have found non-cost attributes important in the decisions made by industrialists (e.g. see Button, 1982) and this is likely to have increased in recent years as the composition of industry has changed.

The studies must also involve making usually implicit but occasionally explicit assumptions regarding the state of affairs that would have emerged if the transport improvements had not been made — they must specify some counterfactual situation. Botham (1983) has suggested that this is not always done very sensibly. In the UK context, for instance, he feels that the congestion that would have built up on the existing network without enhanced transport facilities would have restricted their growth and led to a much greater geographical dispersal in economic activity than has been allowed for in most academic studies. His empirical work leads him to conclude, therefore, that 'the road programme [in the UK] has encouraged spatial concentration'.

A final point of importance is that many of the views of academics and policy-makers we have reviewed were formed during a prolonged period of comparatively full employment which continued throughout the thirty-year period after the Second World War. It is not surprising, therefore, that a succession of both statistical and interview studies (e.g. the latter include those by Fischer (1971) in West Germany and Keeble (1976) in the UK) of factors influencing industrial location found considerations such as labour availability as dominating transport considerations. The macro-economic situation has now changed (possibly permanently) and labour constraints, except in some highly skilled segments of the market, are certainly nowhere near as strong. In the more rigorously competitive environment which has emerged it is quite possible that non-labour factor costs, including transport, are playing a more important role in determining industrial location.

Before moving on to examine the probable future links between transport and development policy, some brief comment on one further dimension of previous attitudes requires comment. This concerns the role transport is seen to play at the micro-level in stimulating the development (or, more often, redevelopment) of specific areas within major cities. Town planners have long viewed transport improvements as a potent instrument of revitalisation policies. The most recent manifestations of this have been the attempts in North America and Europe to stimulate activity in core urban areas by both subsidising public transport and investing in new infrastructure (see also Chapter 3). The *Policy for the Inner Cities* policy statement in the UK made in late 1977 typifies this view when making specific reference to the notion that 'Commerce and industry in inner areas need to be served by transport conveniently and efficiently' and points to the need for local authorities 'to give weight to the implications for local firms when designing traffic management schemes to improve access for central traffic, to ensure efficient loading and to provide adequate and convenient parking'.

The actual power of transport policy to reverse trends towards decentralisation, unless imposed in extreme forms, is debatable. Empirical work is particularly difficult to collect in part because of micro-units involved but also because of the diverse range of factors which have varied in recent years and whose effects it is impossible accurately to isolate from those stemming from transport considerations. Even at the abstract level, there is disagreement about the potential effects of transport provision on urban land-use (see Button, 1982). (Recent advances in urban theory — e.g. the so-called 'New Urban Economics' — have introduced a new rigour into the analysis of this topic but only, to date at least, at the expense of reduced realism.)

Future Policy Trends

The preceding sections have, in general, been couched in terms which might suggest that policy-makers have pursued strict economic objectives when looking at the role transport may play in economic development in its various forms. The theory may not always have been clear but at least the policy-makers have attempted to adopt measures consistent with the prevailing economic wisdom. In practice, this is unlikely to have been entirely true. As we have seen in Chapter 4 when dealing with regulation, the motivations underlying specific actions often stem from less pure influences. There are also practical and, moving away from the specifics of the previous section, much broader political considerations which can play a role. Our concern here is in part, therefore, to look at the importance of these non-economic (at least, in the traditional sense of the meaning of 'economic') factors and to examine the part they are likely to play alongside the strict economic development criteria of transport policy-makers in the future.

The convenient thing about transport and, specifically, transport infrastructure, as we pointed out in previous sections, is that it represents something people can see or at least readily appreciate. Also there are popularly held views, whatever the empirical evidence sometimes shows to the contrary, that transport is important in encouraging economic growth and influencing the locations of economic activities. It, therefore, represents a 'policy instrument' which can be readily identified by politicians and which, in the case of infrastructure provision, is visible. Equally, it represents something which politicians can sell to voters or aid-giving agencies. Grants to industry, tax relief, retraining of labour, guarantees for loans and most other forms of development and spatial policy instruments seldom have this immediate attribute. This is particularly so if there is a perception that public funds may go directly into private investments. To outside agencies providing development aid, transport schemes in the Third World have the further advantage that expenditures are relatively easily monitored and physical results can be observed. There is, thus, some check that funds are going towards the desired end. Transport is also seen to have the secondary effects of bringing distant communities into the mainstream of twentieth-century life and offer the chance of extending social and educational services to them. Improved transport is also often perceived as a means of improving the government's administration over its territories. Thus, even if better transport is not the most effective way to stimulate growth in all cases, at least there is the knowledge that the resources are not being diverted entirely from the wider development programme.

This theme of what is often called 'transparency' has in fact been formalised to a large extent by the EEC in its Common Regional Policy. It, for example, goes a long way to explaining why much regional aid goes on transport infrastructure improvements. The main concern of EEC policy-makers is that they can monitor the way regional funds are used (outwardly economic justifications are advanced for this but in reality there are also often dominant political concerns about equality in the allocation process). Capital projects and one-off expenditures can usefully be converted to a common capital equivalent but longer term, on-going operating subsidies and less direct expenditures are not so easily handled. Transport schemes normally fit very conveniently within the criteria set down in Brussels.

There may be, however, longer term problems associated with an excessive pre-occupation on the capital elements of aid. While not a serious difficulty within a grouping of relatively prosperous countries such as the EEC, the concern with capital equipments can often lead to inadequate funds for maintenance (in fact the US Inter-State Highway Programme encountered this problem as Federal monies were aimed mainly at construction costs, not recurrent outlays). In the context of Third World countries, lack of continuous funding may result in neglect of expensive infrastructure.

There is a further sense that certain types of transport developments are now often preferred, especially in Third World aid programmes. Rural road building can be carried out using local labour and, generally, much locally produced material. This is important in many countries where unemployment (or, in rural areas at least, underemployment) is endemic and where foreign exchange is particularly precious. There is, therefore, an important indirect effect, both through local multiplier effects generated by the construction expenditure itself and the employment created, and through the ability of the Third World countries to redirect their foreign exchange holdings to the purchase of other items. These factors tend to offset to some extent the proclivity of developed countries to provide the actual planning expertise (a fact mentioned earlier in this chapter).

Whatever imperfections may result from the almost inherent bias towards transport schemes in development programmes these varieties of forces create, it is unlikely that in the foreseeable future they will be overcome. We have already pointed to some efforts by donor agencies to ensure that funds delegated to transport are more effectively used (e.g. more efficient cost-effectiveness or cost-benefit analysis may be devised especially with regard to on-going maintenance funding) but it is unlikely that the *overall* position of transport in the league table of perceived investment priorities will slip.

It is also to be hoped that these efforts to develop new allocative techniques circumvent the wider, geographical problems of distributing transport aid. At present aid is concentrated in a relatively small number of countries — in 1980, for example, World Bank and International Development Association aid funded transport schemes in 21 countries but five of them accounted for half the funds available. Further, many of the countries gaining (e.g. the five cited above were Brazil, Romania, Nigeria, Yugoslavia and the Philippines — see Owen, 1984) were middle-income nations rather than those in greatest need. Allocation criteria, therefore, need to make appropriate adjustments for the plight of nations requiring aid but which fall short of standard economic criteria.

Improved international transport has, as we have seen, been one of the factors leading to the growth of economic unions and free market areas of one kind or another. While such unions have as a primary objective the enhancement of members' economic growth they also, in part for pragmatic but also for social reasons, have been concerned to limit any serious disparities in growth rates between member states or sub-regions within the union.

We have already pointed to some of the reasons why regional policy within the EEC, for example, has put resources into transport infrastructure for this sort of reason. While there are likely to be strong administrative forces pushing for the retention of this sort of approach in the future, there are also likely to be growing pressures for deviations from it. Remaining with the EEC, but accepting also that similar conflicts are also probably going to emerge elsewhere, these problems are most likely to develop where the policies of an individual state deviate from the position of the group as a whole. In the UK the idea of regional policy aimed at reducing inequalities has given way to industrial policies concerned with efficiency. The role of transport policy is not seen as central to the latter other than requiring a reactive element to meet industry's needs. The Commission of the EEC in contrast has, since the late 1970s, been committed to developing regional policies with a substantial equity objective and, in particular, with an emphasis on linking disadvantaged peripheral areas of the Community with the more economically advanced core areas. In this sense substantial parts (if not the whole) of Britain are peripheral to the core of Europe and are eligible for regional aid. The paradox is, therefore, that the British government has a specific political interest in ensuring the continued development of this EEC policy approach.

The paradox is heightened because the European Commission contends that transport infrastructure and spatial policy are integrally related. For example in 1979 it argued,

> the less favoured regions must have an internal network of communications appropriate to their present and future needs . . . they must be opened up and linked to the main centres of the Community by rapid modern routes to reduce, as far as possible, the handicap of distance.

To date the Commission has required a member state to justify its selection of projects with reference to that member's domestic expenditure programmes and the objectives which these are designed to achieve. The Commission has then been 'invited' to participate in the financing of some or all of this expenditure. The United Kingdom government has translated this requirement into the preparation and submission of Regional Programmes, which include transport infrastructure projects within the Assisted Areas (or as they have been referred to — the Areas for Expansion). More specifically, this has incorporated projected expenditure plans for the provision of road links to these areas.

The objectives of these parts of the expenditure plans have been related to the selection of individual road schemes. For example, for 1981 these were: to improve links between the peripheral regions and the rest of the United Kingdom; to improve communications between peripheral regions and ports serving traffic with other member states; and to relieve bottlenecks on major international routes. These objectives have in turn been placed in a broader policy context by the Department of the Environment, i.e.:

> The programme will bring substantial benefits . . . It reflects the high priority attached by the Government to schemes which will help the economy . . . The special Community interests . . . are principally greater convergence and integration of the economies of the Community and improved communications between Member States to promote economic and social integration.

In January 1982 the European Commission's grant to assist the United Kingdom's public works programme was boosted by an additional £316 million, as part of the Community's agreement to reduce further the United Kingdom's overall contribution to the European budget. About £100 million of this additional grant contributed to the UK road-building programme, the remainder going to other infrastructural projects — rail modernisation, water and sewage facilities, land reclamation, advance factories, and telecommunications.

The policy problem with which the European Commission has been grappling concerns the principles of allocation which should govern the spatial distribution of these infrastructure grants.

Historically, the bulk have been awarded for specific transport investment projects within the guidelines provided by a member state's declared and Community-approved regional development programme. These programmes are administered essentially by the Commission through their European Regional Development Fund (ERDF). By definition, the criteria for disbursement under-pinning this Europe-wide fund are primarily spatial and require the submission and approval of a nation's general policies for regional development together with a location-specific programme. From a political and administrative point of view then, links between transport infrastructure and spatial policy have been central to the operation of a significant part of the European Community's budgetary process.

A problem for the Commission has been how to justify the allocation of Community funds in general to specific infrastructure projects of member states within the rather nebulous structure of the ERDF. This problem has been compounded further because of difficulties in arriving at a consistent basis on which to evaluate both specific infrastructure projects and a member state's develop-ment programme.

The approach which the Commission has most recently adopted appears to be based on a policy which implies that, when con-fronted with spatially disadvantaged areas, equity considerations should transcend efficiency arguments. Investment in proposed transport infrastructure projects in peripheral regions should be evaluated in such a way that equity considerations are elevated to a central position. The Commission currently argues that evaluation exercises should be based on one or both of two approaches: either by using a conventional form of cost-benefit analysis, incorporating allowances for the social benefits which might accrue to a particular peripheral region as a result of infrastructure investment, or a 'Community evaluation' which is a more qualitative and ordinal form of evaluation. The criteria are based on whether or not particular project, together with the programme of which it forms a part, contributes to the objectives of the Community, especially those relating to assisting the peripheral regions (which includes whole or substantial parts of member states, such as Eire, Greece and Italy as well as the United Kingdom). It is now a Community requirement that if a member state's application for assistance from the ERDF is to be considered by the Commission, then that application must include the results of such an evaluation exercise as well as a detailed impact statement.

It would appear then that the contemporary British government's strategy for transport, location and spatial policy differs significantly from that emerging from the Commission of the European Community. Furthermore, that there seems to be little or no room for compromise. But it should not be forgotten that the Community's evolving strategy is itself tentative and provisional and quite capable of further change. Inevitably, considerable effort will go into seeking compromise in the next few years but it seems unlikely that anything other than a series of *ad hoc* agreements will emerge. Whatever does emerge, it is clear that the tensions between equity and efficiency arguments will not disappear, neither can they simply be assumed away. Historically, the state has always had and continues to have a major responsibility for the level and direction of investment in transport infrastructure and operations. This includes resolving issues about where that investment is to take place and why, which presupposes some type of spatial policy, however crude or simplistic that might be.

References

Anderson, D. L. (1983), 'Your Company's Logistics Management: An Asset or a Liability in the 1980s?' *Transportation Review*, Winter, p. 119.

Botham, R. (1983), 'The Road Programme and Regional Development: The Problem of the Counterfactual' in K. J. Button and D. Gillingwater (eds), *Transport, Location and Spatial Policy* (Gower, Aldershot).

Button, K. J. (1982), *Transport Economics* (Heinemann, London).

Chisholm, M. (1971), 'Freight Transport Costs, Industrial Location and Regional Development', in M. Chisholm and G. Manners (eds), *Spatial Problems of the British Economy* (Cambridge University Press, Cambridge).

Department of Transport (1978), *Report of the Advisory Committee on Trunk Road Assessment* (HMSO, London).

Department of Transport (1980), *Report of the Inquiry into Lorries, People and the Environment* (HMSO, London).

Edwards, S. L. (1970), 'Transport Costs in British Industry', *Journal of Transport Economics and Policy*, vol. 4, pp. 65–83.

Fischer, L. (1971), *Die Berucksichtignung Raumordnungs Politiseler Zeilsetviengen in de Verkehrsplanung* (Berlin).

Fogel, R. W. (1964), *Railroads and American Economic Growth* (Johns Hopkins University Press, Baltimore).

Gudgin, G. (1978), *Industrial Location Processes and Regional Employment Growth* (Saxon House, Farnborough).

Gwilliam, K. M. (1979), 'Transport Infrastructure Investments and Regional Development' in J. Bowers (ed.), *Inflation, Development and Integration — Essays in Honour of A. J. Brown* (Leeds University Press, Leeds).

Judge, E. J. (1983), 'Regional Issues and Transport Infrastructure: Some reflections of the effect of the Lancashire–Yorkshire Motorway', in K. J. Button and D. Gillingwater (eds), *Transport, Location and Spatial Policy* (Gower, Aldershot).

Keeble, D. E. (1976), *Industrial Location and Planning in the United Kingdom* (Methuen, London).

Owen, W. (1984), 'Transportation and World Development', *International Journal of Transport Economics*, vol. 11, pp. 123–33.

Rostow, W. W. (1960), *The Stages of Economic Growth* (Cambridge University Press, Cambridge).

Sharp, C. (1980), 'Transport and Regional Development — with special Reference to Britain', *Transport Policy and Decision Making*, vol. 1, pp. 1–11.

Starkie, D. (1982), *The Motorway Age: Road and Traffic Policies in Post-War Britain* (Pergamon, London).

Wilson, G. W. (1966), 'Towards a Theory of Transport and Development' in G. W. Wilson, B. R. Bergmann, L. V. Hirsch and M. S. Klein (eds), *The Impact of Highway Investment on Development* (Brookings Institute, Washington).

World Bank (1972), *Transportation*, Sector Working Paper (World Bank, Washington).

7   SOCIAL ASPECTS OF TRANSPORT POLICY

Introduction

Transport is rarely if ever used or provided for its own sake. It is a means to achieve objectives associated with everyday life — for example, getting to and from work, shopping, etc. It reflects a basic division which has come to characterise industrial societies — the increasing spatial separation and fragmentation of activities. The demise of the local store or corner shop is one such example; the concentration of major public sector investments in, for example, health care in the form of very large, all purpose hospitals, is another. These trends, combined with the increasing tendency for people's homes to be located in residential areas on the periphery of urban areas, create spill-overs and consequences for transport policy.

This chapter explores the implications of these trends: firstly, by presenting in general terms how transport policy has evolved and followed these sorts of changes; secondly, by presenting the evidence for these changes, and the ways in which transport provision has also changed; and finally, by examining the policy mechanisms which have been deployed to cope with these changes. It concludes with an assessment of the future implications of these trends for transport policy.

This chapter does not pretend to offer a complete review of social and environmental aspects of transport policy. It should be seen as selective in two respects. First, and in line with the approach of the book, the emphasis is placed on trying to identify the principal trends at work in the complex equation between transport provision and transport policy. The emphasis in this chapter is on who uses transport and why, and how these people-centred questions fit into contemporary transport policy. Second, much of the evidence is based on the UK and US experience, and with respect to passenger transport operations. Nevertheless, the implications to be drawn from this partial analysis are likely to be similar for other industrial and industrialising countries, and for other modes.

The Trends

In the recent past, particularly in the 1960s, the key word in transport policy was 'mobility' — creating and sustaining the conditions deemed appropriate to ensure that people could move from one place to another in as relatively unconstrained a way as possible. The visible manifestations of this policy were large road-building and road-improvement programmes linking urban areas, the creation of bus priority schemes in congested central areas of major towns, and investment in public mass transport systems. For example, in the United Kingdom at this time, we witnessed the trunk road improvement programme, the national motorway programme, pedestrianisation of streets, and the creation of a National Bus Company and Passenger Transport Executives. This was all part of a 'total approach' aimed at improving mobility, particularly in the light of forecasts which suggested that private car ownership would reach saturation levels before the end of the century.

This vision of a country being swamped by the demands of private car owners to pursue mobility as part of their personal freedom found its way into two rather contrasting responses. In the public sector the key response was to undertake large planning exercises which became known as 'capacity studies'. The assumption about saturation car ownership levels was in part based on a further assumption about changing demographic patterns, and in particular population forecasts which predicted very large increases. The policy response was therefore based on a supply-led strategy, based on an acceptance of the inevitable growth in population. The policy problem was perceived as one of answering questions about where this population should be located, and how public sector resources could best be marshalled to face increased pressures of demands.

The resultant quest for improving mobility through capital investment in infrastructure and reorganising public transport provision, through a process of rational public planning, was accompanied by a movement dedicated to stopping it in its tracks. Labelled as reactionary environmentalists who wished to halt progress, the environmental movement gathered increasing momentum throughout the 1960s and 1970s, and with it, increasingly influential support (see also Chapter 3). Conflicts between the two opposing movements — those supporting rational

public planning against those advocating an alternative rationality
— became increasingly frequent. One celebrated instance in the
UK saw the headmaster of Winchester College — a renowned
private fee-paying public school — being forcibly ejected from a
public inquiry into a proposal for extending a nearby motorway. In
the United States, the organised protest against major road construction in Boston in the late 1960s eventually proved irresistible,
with the Governor of Massachusetts being forced to call a
moratorium.

This is not the place to explore all arguments for such differing
views. However, in transport policy terms, a major source of
controversy between the two movements centred around 'public
interest' arguments. Overgeneralising a very complex period, the
rational public planning proponents agreed that it was in everyone's interest to pursue improvements in mobility as a major
policy objective. Although there were often clear disagreements
about the means to achieve this objective, for example, about the
scale and programming of priorities for investment, there was a
clear consensus that this approach was both legitimate and in the
public interest.

The disparate groups which made up the environmental
movement were equally in agreement that such a policy was not in
the public interest. In developing their equally plausible
rationality, public interestism was at the centre. Their shared
starting point addressed the likely consequences which would arise
were such a policy to be pursued, and focused on four main
problem areas. Firstly, the presumption that it was in everyone's
interest was challenged on strictly environmental grounds, particularly with regard to pollution effects and the massive land-take
associated with major road building projects. Technical evidence
was amassed and presented at public meetings and inquiries.

Secondly, the policy was challenged on social grounds, and in
particular by the presentation of evidence showing that new roads
would be constructed largely at the expense of those living in
urban areas on low incomes, where the housing stock was fairly
old, and where levels of car ownership were significantly below
'average'. This distributional issue — the fact that those who
would benefit most would suffer least — introduced a class-based
dimension to the public-interest argument.

Thirdly, and mainly in the UK, the policy was opposed on
procedural grounds. Increasingly, those opposing the road-

building programme could only fight each scheme on an individual, piecemeal basis. Despite unsuccessful attempts to challenge the overall national policy in the courts, the only way open for protestors to make their case was in the local public inquiry. This quasi-judicial forum, chaired by an inspector appointed by the same government department that was responsible for implementing the policy, came in for increasing criticism. At inquiry after inquiry, successive inspectors would not permit discussion of any policy issue: they consciously and expeditiously steered all controversy into the merits and demerits of alternative schemes. In effect this meant that the inquiry only seriously addressed the strengths and weaknesses of alternative highway lines. As a result, not only were such inquiries becoming increasingly oblique in their treatment of different arguments based on an alternative rationality of the public interest, but they were also becoming increasingly protracted. In a recent study by the National Economic Development Council, it estimated that on average it was taking ten to twelve years to complete a major road-building project, with much of this time being taken up with two or more public inquiries. The decisions — or rather series of non-decisions — relating to the choice of a third international gateway airport for London is one of the clearest examples of this protracted procrastination. It has taken about thirty years to arrive at a final decision to expand Stansted Airport.

It is, however, the fourth challenge to the mobility policy which is of greatest consequence. Following the débâcles which surrounded the rational public planning approach, and in particular the analysis of distributional consequences highlighted earlier, a new movement began to emerge. There became an increasing awareness that, on the one hand, the assumptions about an 'inevitable' growth in population which underpinned rational policy was questionable and, on the other, that mobility was only a means to an end in transport terms. What really mattered was not how people could improve their mobility, but rather why people wished to travel. If there was a precise moment when the rational public planning approach lost its credibility, then it has to do with the constellation of events in the period 1973 to 1976 — a combination of a major downturn in world trade leading to recession, the Arab states' oil embargo, and an awareness that the share of public sector resources was under real strain.

This shift in emphasis away from a mobility based policy and towards accessibility as a policy objective should not be under-estimated. In the UK, one movement was particularly influential — the self-styled Independent Commission on Transport's report entitled *Changing Directions* (1974). At the same time, Political and Economic Planning (PEP) had initiated its own compre-hensive study of personal mobility and transport policy, again with an emphasis on a critique of mobility based policies (Hillman, Henderson and Whalley, 1973). At a similar time in Europe, the Organization for Economic Cooperation and Development began to focus on similar issues, but more from an environmental viewpoint (OECD, 1973). In the United States, the Brookings Institute published an equally influential study, stressing a policy argument based upon accessibility — *The Accessible City* (Owen, 1972).

The common theme running throughout these studies was that transport policies were too concerned with the supply side. They argued that, as a consequence, two factors came to dominate thinking behind policy: first, that at almost whatever the cost, the expected growth in private vehicle ownership and use had to be accommodated. As a result, transport policy had become effectively a vehicle-orientated policy; and secondly, this reactive approach ignored the other, arguably more important, side of the equation — the demand for transport. This is not to say that the demand side had never been a major feature of transport policy. In this sense demand was seen as the best-guess forecast of the number of vehicles expected to be travelling around at some predetermined future date, together with its modal split (i.e. the proportion of cars, buses, trains, lorries, etc.). This statistical modelling exercise was complemented by, but divorced from, another type of demand-based work — the allocation of road space through the application of economic principles of efficiency. The civil engineering dominance of transport policy was reflected in the large-scale, supply-side and supply-led 'land use–transportation studies' so typical of the era. The economic content of the demand-side of policy focused very much on the issue of road pricing as a way of allocating usage within a framework provided by the supply-led programme (see Chapter 3).

The new accessibility approach challenged this way of con-structing a transport policy. It proved to be a remarkably effective assault. By the end of the 1970s the strategy behind transport policy reflected the sort of changes noted before. Perhaps the most significant effect was the redrawing of the institutional arrangements responsible for the planning and programming of its implementation. In the UK, the whole approach to preparing and funding transport schemes moved away from the blueprint strategy of preparing a long-term physical development plan. In its place a more present-oriented programme-based approach was adopted, through a mechanism called 'transport policies and programmes', and for public transport provision, 'public transport plans'. In the United States a similar change took place — away from the long-term, one-shot land use–transportation study approach and towards a more pragmatic approach called 'transport systems management'. Similar events occurred in many other industrialised countries, such as the Netherlands, Sweden, West Germany and Canada.

It would be a mistake, however, to think that these changes were brought about solely as a result of a rethink about the orientation and implementation of transport policy. Of equal importance were the changes which were occurring within transport provision. On the one hand, private modes of transport were undeniably in the ascendancy, both in absolute and relative terms. The traditional conventional forms of public transport were experiencing a demise, both with respect to patronage and levels of service.

On the other hand, it was clear that not all demands for transport provision were being met. Large tracts of industrialised countries were becoming characterised by low car ownership and use, and no correspondingly adequate provision of public transport services. These accessibility gaps were particularly evident in rural areas and within central zones of metropolitan areas. At the same time, a large number of vulnerable social groups — such as the elderly, handicapped persons, and young children — were effectively experiencing mobility deprivation not of their own choosing.

These effects were in turn being exacerbated by two further contradictory trends — firstly, a continuation of the steady geographical dispersal of economic activities; and secondly, the pursuit of policies in other public sector agencies, such as health care, resulting in the further concentration of public services in central, often large city, locations.

Whether these trends will continue is a matter for conjecture. But the issues which became highlighted in the accessibility studies of the 1970s are demonstrably still present more than ten years on. As one recent UK study puts it:

> The trends that argue too strongly for car and freight vehicles cast shadows over the prospects for rail and bus transport. [The Department of Transport's] forecasts have nothing to say about them or the groups who may wish to use them. Indeed, they do not even tell us the numbers of households, still less people, who will not be able to rely on the use of a car. Nor do they say anything about the nature of the trips that such households might be trying to make. Yet, there are indications that the difficulties facing those without cars will be greater than they are now (Harrison and Gretton, 1985).

The Evidence

To place a people-centred analysis in perspective, Table 7.1 presents information on personal mobility over a period from the 1950s to the 1970s for Britain and the USA.

In the USA, car-based personal mobility increased by 196 per cent in the twenty-five years from 1950 in terms of distance travelled. In Britain the corresponding figure was 440 per cent, although calculated on a slightly different basis and over a later time period. Nevertheless, the picture is the same: a remarkable increase in the personal mobility of the two populations. In 1975 the average US citizen travelled over 27 miles each day, a 180 per cent increase in twenty-five years. In Britain, the average citizen travelled over 15 miles each day in 1977, an increase of 222 per cent since 1954.

There are, however, problems with this analysis. First of all, it has little to say about who has been travelling and why the distance travelled has increased. Also it is based entirely on averages and thus it says nothing about either the distribution of trips nor the ranges involved. In other words, the nature and characteristics of the persons involved are concealed in a generalised picture.

It was in part to counter this kind of mobility-based style of analysis that led to a move to introduce people into transport policy, through the mediating principle of accessibility. In the most comprehensive study of its time in Britain, Hillman, *et al.* (1976) tried to establish the research-based ground rules for a people-centred approach. Their research was designed 'to provide a comprehensive view of the ways in which mobility and travel behaviour vary between different people living in areas with contrasting physical characteristics'. The result was to argue a case for accessibility-based transport policy, which incorporated, for the first time, detailed social survey results in five quite contrasting localities.

Their starting point was to scrutinise closely the kind of mobility- and modal-based studies noted earlier. Of the major defects which they identified, the two most salient were to question (i) the validity of using 'the household' as the basic unit of survey in mobility studies, and (ii) the selectivity which was identified in these studies' choice of particular groups of people, types of journey and methods of travel to survey.

Emphasis on the household introduces a distorting effect in social survey-based studies. On the one hand, the roles, preferences and mobility of individuals within households usually vary widely; and on the other, as there is often a relationship between each individual's travel needs, mobility and travel behaviour, any studies which focus on 'the household' as a whole tend to be misleading.

The selection of particular social groups, journey types and mode — it was argued that: 'in studies where individuals have been recorded, attention has been focused on the head of the household; this person is usually male and often the only person capable of driving in those households which are car owning'. The implication is that other individuals making independent trips are excluded, leading to an under-representation of, for example, trips by women and children.

Their second accusation of partiality in mobility studies was levelled against the selection of journey type, and in particular the almost exclusive attention paid to 'the journey to work' — as they note this is the only journey for which national census returns have been taken. Whilst concentration on this type of journey is understandable, it is difficult to see why it should be given exclusive consideration. As they state:

the *traffic* problems of the rush hour are only seen to be of consequence because limitations of road capacity necessitate restraint on car use in urban areas. Yet the *social and environmental* problems caused by traffic may be more

Table 7.1: Personal Mobility in Britain and the USA

	Miles travelled per person per year				
United States		*1950*		*1975*	*1975 as a percentage of 1950*
Car		4,567		8,987	196
Surface public transport					
Intercity bus	172		132		77
Intercity train	212		47		22
Urban transit	570		164		29
		954		343	36
Airplane		66		686	1,039
		5,587		10,016	180
Miles per person per day		15.3		27.4	180
Britain					*1977 as a percentage of 1954*
		1954		*1977*	
Car, taxi, motor cycle		961		4,234	440
Surface public transport					
Bus	1,024		606		59
Train	493		389		79
		1,517		995	66
Airplane		32		365	1,141
		2,510		5,594	222
Miles per person per day		6.9		15.3	222

Source: Adams (1981).

disturbing at other times; for instance, vehicles on a street impinge on its traditional function of being a locus for social activity during the day time; and their noise can discourage people from opening windows in their houses in the evenings and at weekends [original emphases].

The survey evidence in the study highlights the complexity of transport use. Beginning with an examination of car ownership and use in five quite contrasting localities, they proceed to explore 'the realities' of daily travel with respect to the differing circumstances of particular social groups (adults, teenagers, schoolchildren, young women with children, and elderly persons). Table 7.2 presents their findings on car ownership and licence-holding by the five localities. It shows wide variations according to where people live. The lowest level of car ownership was in inner London, the highest being in a new town. About a third of all households in a village or small town did not own a car and in these two localities, between 39 and 45 per cent of all adult respondents had no driving licence.

Table 7.2: Car Ownership and Personal Licence Holding in the UK

Cars in household	Village %	Small town %	New town %	City suburb %	Inner London %
None	33	29	36	54	57
One	49	59	60	40	38
Two or more	18	12	4	6	5
Driving licences					
No driving licence	45	39	49	61	66
Provisional licence	4	3	6	4	3
Full licence	51	59	44	35	31

Source: Hillman *et al.* (1976).

Table 7.3: Access to Cars

	Household has	Respondent has	Village %	Small town %	New town %	City suburb %	Inner London %
Men (all ages)							
Level 1	Car	Licence	67	70	64	54	40
Level 2	Car	No licence	6	7	3	1	7
Level 3	No car	Licence	8	7	10	10	9
Level 4	No car	No licence	20	16	24	35	44
Women (all ages)							
Level 1	Car	Licence	33	36	14	8	13
Level 2	Car	No licence	31	27	45	30	25
Level 3	No car	Licence	2	3	5	2	1
Level 4	No car	No licence	34	34	37	60	61
Elderly persons							
Level 1	Car	Licence	24	13	18	3	4
Level 2	Car	No licence	18	13	6	3	4
Level 3	No car	Licence	–	3	6	3	–
Level 4	No car	No licence	58	70	71	91	91

Source: Hillman *et al.* (1976).

Owning a car is, however, no guarantee that it is available for those members of a household capable of using it. Table 7.3 presents evidence which examines the incidence of car use and availability, according to levels of access. The highest — level 1 — is enjoyed by those holding a licence and living in a car owning household. The lowest — level 4 — have neither licences nor cars.

Three striking features emerge: firstly, that irrespective of locality, elderly persons are most likely to be on the lowest level with neither car nor driving licence; secondly, that over 90 per cent of elderly persons in city suburbs and inner-city localities are so disadvantaged; and thirdly, that there is a significant difference between the access of men and women to cars.

The implications for non-car based forms of transport provision also emerge in Table 7.3. For example, 46 per cent of men in city suburbs either had no access to a car or no licence, rising to 60 per cent in inner-city areas. For women, this picture is intensified: 92 per cent having no access and/or no licence in city suburbs and 87 per cent in inner-city localities. For elderly persons the corresponding figures were 97 and 96 per cent.

The method of travel — also examined — used most frequently was walking. This represented the majority of all trips in villages, falling to just over a third in small towns. The data also shows the clear

Table 7.4: Journey Purpose and Method of Travel

Type of journey		Village %	Small town %	New town %	City suburb %	Inner London %
Shopping	Walk	64	58	67	65	73
	Bus	6	12	11	16	19
	Car	29	30	23	19	8
Work	Walk	35	27	13	17	17
	Bus	3	13	20	35	59
	Car	63	60	67	49	24
Social	Walk	58	44	48	48	44
	Bus	5	7	6	11	24
	Car	37	49	46	40	31
Leisure	Walk	47	30	40	41	57
	Bus	6	8	8	14	18
	Car	47	62	52	44	25

Note: 'Bus' includes rail and underground services.

Source: Hillman *et al.* (1976).

locality differences in the use made of public transport — from as low as 5 per cent in villages rising to 33 per cent in inner London.

What also emerges is that travel method significantly depends upon the purpose of the journey. Table 7.4 presents findings for this relationship.

The importance of walking for shopping trips is clear. For work journeys, however, a more complex picture becomes apparent. For those working in a village locality, over a third walked to work, and almost two-thirds used a car. This can be contrasted with those working in inner-city localities, where the majority used public transport. Reliance on the car for work journeys is very high for the village, small town and new town, falling considerably for the city suburb and inner-city localities. In contrast, reliance on public transport is strongest in inner city and suburban locations.

The summary of the study presented thus far has focused attention on adult travel only. To illustrate their concern for other large categories of personal travel, Hillman *et al.* undertook further studies into the particular transport problems of children, the elderly, and those using public transport.

Table 7.5 presents findings for children's independent travel to a range of activities. By far the largest method of travel is walking, irrespective of journey purpose. The exception is travel to leisure activities, where the dominant mode is the bus. Travel by car is obviously excluded, since children cannot legally drive. The last column in the table shows the dependence of children on others. Although not included in Table 7.5, Hillman *et al.* also examined the differences in accessibility according to the five localities. Children living in suburban localities appeared to have the most independence,

Table 7.5: Children's Independent Travel

	Child travels alone:			Child does not travel alone
	On foot (%)	By bus (%)	By cycle (%)	
Shops	86	20	32	6
Playgrounds/parks	79	12	32	10
School	81	9	1	14
Social visits	71	22	34	15
Lessons/clubs etc.	53	22	16	32
Leisure	35	40	11	32

Source: Hillman *et al.* (1976).

whereas those in the inner city were severely constrained, with cycling particularly low.

A particularly interesting part of the Hillman *et al.* study concerned the analysis of the mobility problems facing elderly people. Of the elderly people surveyed, 41 per cent said that they experienced problems when walking. Of these, the three most frequently cited were: changes of level, involving hills and/or ramps; uneven pavements; and the speed and/or volume of traffic.

For those users of public transport, the survey attempted to identify the type and frequency of problems for three social groups — the elderly, young women travelling with children, and women travelling alone. Those with the greatest number of problems were, perhaps surprisingly, young women travelling with children, although as Table 7.6 shows, many of the specific problems were common to all groups.

Table 7.6: Problems with Use of Public Transport

Problems cited	Elderly persons	Young women travelling with children youngest child is:		Women travelling alone
		Under three	Three or over	
	(%)	(%)	(%)	(%)
Infrequency	37	16	36	46
Unreliability	37	} 9	} 33	} 36
Waiting	22			
Effort	20	62	20	8
No space for push-chair	n/a	34	12	n/a
Unsuitable routes	15	16	6	8
Stop too far away	11	2	7	5
No bus shelter	11	8	2	5
High fares	8	11	17	17
No conductor	6	31	9	3
Conductor doesn't help	1	39	11	3

Source: Hillman *et al.* (1976).

The general picture which the Hillman *et al.* study has built up suggests that personal mobility problems are both more complex and localised than previous transport studies and policies had presupposed. Other studies have borne this out. As part of a broader project on rural transport in Norfolk, England, Moseley (1979) highlighted the condition of 'multiple mobility deprivation' — a condition he characterised by the incidence of nine factors: no car, no local bus service, no local shop, someone in the household who was physically handicapped, low income, no telephone, no public or private telephone available within five minutes walk, no bicycle, and no refrigerator. Of those elderly households surveyed without a car, 67 per cent suffered at least three more factors from the list, and 22 per cent suffered at least five more.

Table 7.7: Social Trips by Elderly Persons

| Journey purpose | Elderly housholds | |
in previous month	With car(%)	Without car(%)
Library	41	19
Public house	40	22
Cinema/theatre	9	2
Hospital visit	17	11
Visiting friends	52	27

Source: Moseley (1976).

Table 7.7 shows the incidence of social trips by elderly households in rural areas, according to car availability. In each case it shows that those without cars experienced significantly less trip-making than those with cars. As Moseley puts it:

All surveys of the mobility and accessibility problems of people living in rural areas point to the elderly being the single most important disadvantaged group. Studies both of how they actually behave and of the opportunities afforded to them by their 'environment' of cars, buses and other services point to the severity of their disadvantage. And yet how many rural plans . . . explicitly include the welfare of the elderly as an objective whose degree of attainment is to be carefully sought and then monitored?

However, mobility deprivation is not restricted to elderly people, even though they may represent the single largest group. A different style of accessibility analysis, associated more with North American research work, focuses attention on access to facilities. A typical example is provided by Wachs and Kumagai (1972) in their study of access to health-care facilities in Los Angeles. They studied four groups of people living in two contrasting localities, those with access to a car and those dependent on public transport. The aim of the study was to identify the health-care opportunities available to these groups, in terms of how many doctors, hospitals and clinics they could gain access to in a given time period. The summary results are presented in Table 7.8.

Table 7.8: Accessibility to Health-Care Facilities in Los Angeles

Facilities	South Central		Bell Gardens	
	By car	By public transport	By car	By public transport
No. of health-care opportunities within 15 minutes travel time:				
General practitioners	335	11	285	18
Hospitals and clinics	40	2	41	0
Total	375	13	326	18
No. of health care opportunities within 30 minutes travel time:				
General practitioners	1534	112	1529	36
Hospitals and clinics	143	14	149	1
Total	1677	126	1678	37

Source: Wachs and Kumagai (1972).

Taking this approach to accessibility demonstrates the unequal opportunities open to those individuals with a car and those reliant on public transport. In both localities car ownership conveys a clear advantage by a factor 23.

Another study conducted in the US focused on problems of mobility and unemployment in rural areas. Maggied (1982) showed that 15 per cent of rural households do not own a car, and 52 per cent own only one. The problem is particularly severe for the rural poor, 57 per cent of whom do not have a car, and for the rural elderly, of whom 45 per cent are car-less. The estimate of 9.6 million officially defined rural poor have substantially less access to public transport than does the rest of the population. These disadvantaged people also have to travel much greater distances to work, as well as obtain essential services. Maggied's study shows that an estimated 2 million rural workers commute on average 25 to 30 miles each way. The dependence on private transport is especially high, since only 315 of the country's 20,000 non-metropolitan authorities have a public transport system.

Maggied's study had a particular policy cutting-edge to it, since its aim was in effect to determine Georgia's statewide transport needs under the Federal government's Section 147 demonstration

Table 7.9: Personal Income, Mobility and Work Activity in the USA

Area	Commuting workers (%)	Unpaved county roads (%)	Family income less than $8000 (%)
United States	19.2	22.6	27.7
Georgia State	26.8	57.0	31.8
Dawson County	45.3	75.7	53.1
Baker County	41.8	71.6	67.5

Source: Maggied (1982).

project programme of the 1973 Federal Aid Highway Act. It was designed to consider the possibility that limited personal mobility impedes access to work activities, which in turn determines personal income. He suggests that this mobility/personal income/work-activity relationship is circular and continuous. Table 7.9 presents one of the principal findings. What emerges is 'that more than half again as many workers journey from the home county to work across county lines over more miles of poor roads to receive a substantially lower gross personal income'.

Table 7.10: Bus Trips and Household Income in Sheffield and Rotherham

Household income group		Bus Trips per person in Urban Sheffield/Rotherham	
1972	1981	1972	1981
0–£749	0–£2,704	0.63	0.64
£750–£1,249	£2,705–£4,509	0.76	0.82
£1,250–£1,749	£4,510–£6,314	0.73	0.81
£1,750–£2,499	£6,315–£9,024	0.71	0.84
£2,500–£3,999	£9,025–£14,439	0.71	0.67
>£4,000	>£14,440	0.73	0.54
all incomes		0.71	0.76

Source: Goodwin et al. (1983).

In a recent major British study of the demand for travel, a large-scale survey was conducted of the consumers of transport (Goodwin *et al.*, 1983). Although confined largely to the administrative area of South Yorkshire, focusing particularly on the urbanised areas of Sheffield and Rotherham, the survey tried to assess the impact of subsidies on public transport patronage. Detailed studies were undertaken of the socio-economic charac-teristics of a sample of the total travelling public. What it found is that the single largest group of users were those in full-time employment, although the highest trip rate was associated with those in part-time employment. The next largest group of users were students.

In terms of the relationship between public transport use and income, Table 7.10 presents the study's findings for urbanised areas.

Over the period 1972 to 1981, trip-making by bus increased a little over 7 per cent. Perhaps paradoxically, the lowest rate of trip making was associated with those on the lowest level of earnings, and the group which experienced the greatest increase — over 18 per cent — were in the middle income range.

An analysis of bus trips by age and sex shows that women were the principal users, and particularly those aged between 35 and 64. The largest single group of male users was in the same age band. In terms of changes in trip-making between 1972 and 1981, the age bands experiencing the biggest increases in shares were children (5–14) and young women (15–24). Decreasing shares were associated with all males (15–65+), and middle-aged males and females (35–64).

The relationship between public transport use and car ownership proved particularly illuminating. Over half the households surveyed made daily use of bus transport. Perhaps unsurprisingly, the households making most daily use were those without a car, although 45 per cent of those with one car included regular bus users. This proportion dropped for households with two or more cars, but even then 39 per cent included members of households who were daily bus users.

The implications which flow from all this evidence about personal mobility and accessibility suggests that transport, and particularly the split between private and public transport, is very much more complex than the image which has hitherto characterised transport policy debates. The main points are that

(i) cars are not the universal form of transport which most transport planners have assumed them to be;

(ii) it is not car ownership which matters, but rather individuals' access to its availability;

(iii) public transport is not the most widely used form of transport for those without either ownership or access to a car — walking is the most common form of mobility; and

(iv) for those with lower levels of access to transport, their personal activity patterns are constrained, but are not curtailed.

For groups like the elderly and infirm the problem of providing access and mobility has been increasingly picked up by statutory agencies (like local authority social services departments), voluntary organisations (like local Councils for Voluntary Service) and community groups (such as community transport projects). According to Benwell (1985), these 'special needs' groups cover about 10 per cent of the adult population in the UK, and include not just the elderly but disabled persons, those with mental health conditions, the temporarily disabled, school children and 'children at risk'. If groups like unemployed people are also included, this figure increases to nearer 30 per cent of the population. However, although the total population of the transport disadvantaged is quite large, there is little uniformity with respect to their particular travel needs. For example, what might be the most appropriate transport solution for an individual temporarily disabled and requiring out-patient medical care is clearly quite different from the solution for getting young children to school. The problem from the viewpoint of transport policy is that unless such transport is provided, whole minority groups are deprived of the opportunity to participate in the full range of activities which are regarded as being part of everyday life.

The policy response has tended to take two quite contrasting paths. On the one hand, whole groups of disadvantaged people have been incorporated into legislative schemes whereby their transport needs are deemed meritable. Getting children to and from school, getting people to and from hospital out-patient appointments, and conveying young children at risk to day nurseries are just three examples. In England and Wales, these particular needs are met through the Education Act 1944, the National Health Service Act 1977, and the Children and Young Persons Act 1963. In the United States similar legislation applies, both at state and federal level. For example, the Rehabilitation Act and amendments, the Urban Mass Transportation Act and amendments, and the Older Americans Act. In France, the main piece of enabling legislation, particularly for disabled and elderly persons, is the Outline Act (Orientation of Handicapped Persons) 1975 under the 7th National Plan.

The second path has developed largely in response to the provision of unconventional transport services by non-government agencies. It is in Britain and the United States where this form of transport provision has become in effect a secondary level of public transport. Although the evidence is patchy, because of its very local roots, these services take three main forms: voluntary car services; community minibus schemes; and dial-a-ride services. For example, by 1984 there were over 300 separate community minibus schemes in Britain, and the number of voluntary car schemes are probably at least double that number. In both countries the relationships between voluntary based transport projects and statutory public agencies are now so intertwined that the latter are now heavily dependent upon the former to meet their statutory obligations. As Norman (1977) puts it:

> Voluntary schemes can be used to provide local public transport where commercial operation is uneconomic; to provide specialist transport to health centres and other medical or social services; to carry elderly people to day clubs and day centres; to take elderly visitors to distant hospitals and to meet a host of individual needs. Without it many statutory services could not run at all and many elderly people's lives would hardly be worth living.

In one of the few studies of voluntary transport in Britain, Bailey (1979) examined provision in Birmingham. There he found that the 122 vehicles operating in the area were providing a similar level of social transport as the local authority social services department. Table 7.11 presents his findings by user group. In total, Bailey estimated that about 35 million trips were provided to meet special needs.

Table 7.11: Passenger Trips Provided by Voluntary Transport in Birmingham

	Number	%
Elderly	299,257	27
Physically handicapped	343,591	31
Mentally handicapped	121,919	11
Mentally ill	11,083	1
Children and families	254,923	23
Other	77,585	7
Total	1,108,361	100

Source: Bailey (1979).

It is the community minibus and dial-a-ride schemes which have generated considerable interest in transport policy terms. In

Britain, the United States, Holland, Sweden and West Germany, this style of unconventional transport provision has been seen as providing a complementary type of service, along with conventional public transport and use of private cars. Although the British experience was not seen as successful in its early days of the 1970s, there can be little doubt that it presently represents a major growth sector in public transport provision. Most dial-a-rides and community minibus schemes are community-based and locally controlled. Although a large proportion of capital funding, particularly for vehicles and maintenance, is often provided by public sector agencies, all schemes rely for their operation on the personal commitment of volunteers and a few paid co-ordinators. In London, for example, there will soon be complete coverage, although at present there are 12 projects. Table 7.12 presents some limited evidence on six of these schemes, according to journey purpose.

Table 7.12: Dial-a-Rides and Trip Purpose (percentages)

	Social	*Shopping*	*Health*	*Education*	*Work/ Community Business*	*Inter- mode*	*Other*
Haringey	56.3	24.4	6.9	11.3	0.9	0	0.3
Kensington and Chelsea	52.3	26.7	13.6	4.0	2.8	0.6	0.6
Ealing	46.4	16.3	26.1	1.3	2.6	1.3	5.9
Greenwich	42.5	38.6	2.6	2.6	10.5	2.0	1.3
Westminster	51.5	26.3	6.1	7.1	3.5	1.0	4.5
Hammersmith and Fulham	62.8	18.2	4.1	2.5	5.0	2.5	5.0

Source: National Advisory Unit for Community Transport (1985).

In each scheme, the largest single journey purpose involved social trips, accounting for more than six out of every ten trips in one scheme. Only one project was used for more than a quarter of its total trips for health purposes.

Of the trips for social purposes, in each case about half involved visiting friends and relations, and between 11 and 40 per cent for trips to clubs and groups.

Currently, the UK Department of Transport estimates that it is issuing about 2,000 minibus permits each year, many of these being used for organised transport operations like dial-a-rides. In the United States, this style of transport provision is often the only form of public transport to be found in small towns and deprived areas. At the federal level, the Urban Mass Transportation Administration funds a substantial research and support programme, although in 1977 it was estimated that at least 114 different federal schemes were aimed at the transport disadvantaged, with a combined budget of about \$2.2 billion — an estimated 5.5 per cent of the broader programmes with which these transport schemes were associated. There were in addition an unspecified number of state level schemes. In one state, for example, North Carolina, the Governor in 1978 issued an Order creating a Public Transportation Advisory Council, with one of its objectives being to promote co-ordination between public transport projects and operations. By 1984, 32 such schemes were in operation, of which 18 were operated by a city or county administration, 10 by private

Figure 7.1: The Provision of Public Transport Services

	Fixed-route service	Variable-route service	Contract hire service	Demand-responsive service
1) *Primary* e.g. Public transport	Conventional bus service	Post buses	Private coach and minibus hire	Private taxis
2) *Secondary* e.g. Voluntary transport/ Community transport	◄——— Community buses ———►		Urban community transport	Dial-a-ride and social car services
3) *Tertiary* e.g. Public sector transport	◄— School buses —►	Outpatient ambulance service		Hospital car service
	◄——— Social services transport ———►			

Source: modified from Sutton (1986).

operators (usually taxi firms), 3 by private management companies, and only one by a public service agency. The income ranged from $600 a year to $212,000, with an average of $68,600, reflecting differences in the scale and level of operation, and the area served. The four smallest operations were all privately operated. Although there is little information about patronage, it is clear that the schemes are aimed at elderly persons and handicapped people (Zeitlin, 1984).

In terms of the social aspects of transport policy, the evidence suggests a style of public provision which has three levels offering four contrasting types of service, illustrated in Figure 7.1.

What this figure shows is the complementary nature of public transport provision in contemporary society. The primary level of public transport covers, for example, conventional bus services, the coach market and taxis, as well as rail services and air passenger services. The objective of this level of provision is to meet the demands for personal mobility, where the market mechanism plays a central if regulated role. The third level, what we have called public sector transport, includes those services provided by statutory agencies like local authorities. Here the objective is to meet statutory obligations, usually associated with non-transport policies, such as health care. The concept of meeting effective demand does not appear at this level, the allocative mechanism relying principally on rationing devices.

The edges between the two quite different types of transport provision have become increasingly blurred, particularly so with the arrival of a third force in public transport provision. This level intervenes between conventional demand-centred public transport and special-needs-based transport provision. It operates as more than a gap-filling exercise, since it has arisen largely as a result of the incomplete level of service of the total public passenger transport market vested in the primary and tertiary levels.

Figure 7.1 shows this total market segmented not only by three levels of operation, but also according to the nature of the transport system provided. Conventional transport policy has been locked into arguments about the merits or otherwise of fixed route networks, like buses, versus demand responsive services, such as taxis. Variable route and contract hire operations have largely been left to policies associated with public sector transport provision, transporting individuals from their homes to institutions like day centres and back again. What Figure 7.1 shows is a view of

198 Social Aspects of Transport Policy

the complementary nature of these quite different and contrasting transport systems.

The Policy Mechanisms

As previous chapters have shown, transport accounts for a very large share of resources and contributes a significant amount to national prosperity. In the UK alone, the resource costs of transport in 1983 have been estimated at over £52 billion, and the contribution to gross domestic product was of the order of 17.3 per cent. Road-based transport accounted for 93 per cent of this total. As Plowden (1985) demonstrates, included in this was over £5,000 million of public expenditure, divided into 28 per cent for public passenger transport and over half on road construction and maintenance.

In the US in 1981, the budget of the federal Department of Transportation alone was over $18 billion, with over 45 per cent being allocated to the Federal Highway Administration and almost 17 per cent to the Urban Mass Transportation Administration. If an estimated $38 billion of state level expenditures are added, the combined total comes to $56 billion in one financial year. At state level, about 89 per cent is spent on road construction and maintenance.

A recent study of transport resources in the UK (Harrison and Gretton, 1985) suggests that conservative estimates by principal sector can be made, which attributes £2.1 billion to central government, £3.3 billion to local government, £6.1 billion to public sector organisations (mainly buses and rail), £11.9 billion to private sector organisations (mainly road freight), £10.6 billion to commercial 'own account' operations (mostly road haulage) and £16.4 billion to households (almost exclusively personal cars).

How these expenditures are allocated, and the form which they take, reflect not only the pressures and changes in transport provision as a result of previous rounds of investment decisions and the workings of the market, but also attempt to influence them through the application of transport policy. These general policies are grounded in specific mechanisms which mean that decisions are made and commitments entered into involving detailed capital spending programmes (like new roads) and the funding of specific revenue programmes (like public transport subsidies).

Table 7.13: UK Government Expenditure Plans 1981–2 to 1984–5

	£ million cash 1981–2		Estimated 1984–5	
CENTRAL GOVERNMENT FINANCE				
Department of Transport				
Motorways and trunk roads	639		770	
Subsidies to transport industries	919		1,050	
Ports	109		110	
Other transport services	60		70	
Driver and vehicle licensing	73		90	
Department of Environment				
British Waterways Board	28		35	
Other transport services	22		10	
Department of Trade				
Shipping	31		30	
Civil aviation	19		15	
Scottish Office				
Roads and transport	131		148	
Welsh Office				
Roads and transport	114	143		
		2,145		2,471
LOCAL GOVERNMENT FINANCE				
Capital				
Roads, new construction and improvement	338			
Car parks	15		850	
Public transport investment	250			
Local authority airports	31		35	
Scottish Office	148		168	
Welsh Office	49	61		
		831		1,114
Current				
Roads maintenance	692			
Car parks	–18		1,540	
Road safety etc.	13			
Local authority administration	230			
Passenger transport subsidies				
British Rail	47			
Bus, underground and ferry services	377			
Concessionary fares	157			
Scottish Office	215		244	
Welsh Office	93	116		
		1,806		1,900
Total		4,782		5,485

Source: HM Treasury (1982).

The main vehicles used to deploy and direct these resources are expenditure plans and land-use/transport development plans. These mechanisms are to be found in all countries across the globe, but particularly refined examples are illustrated in the British and North American experiences.

In the UK, the main policy mechanisms for the allocation of transport expenditures are vested in annual five-year 'public expenditure programmes' of central government, and local authority organised annual 'Transport Policies and Programmes'. In the USA, similar mechanisms are at work, but operating at both state and federal levels. The example to be examined here is the 'transport systems management' process.

Table 7.13 shows the way in which transport expenditures are managed by central government in the UK.

At local authority level, these central government estimates provide the guidelines within which the annual 'transport policies and programmes' are prepared, and give a crude indication of the likely expenditures which central government will find acceptable and eventually approve. The link between central government expenditure on transport and local authority programmes is currently a mechanism called 'Transport Supplementary Grant'. This is a bid by each local authority for grant aid from central government. The policy objectives and spending priorities are set by the centre, the localities determine their bids within those constraints, the bids are submitted, and the centre allocates grant accordingly. This process takes place annually, and the justification for grant aid is made within a statutory document called a 'TPP'.

A typical TPP has four sections:
(i)   an introduction, stating the local authority's objectives and policies for transport in its area, together with a summary of the bid and its priorities;
(ii) an analysis of the key policy issues within which the bid is framed — for example, road safety, public transport development and highway maintenance;
(iii) the detailed bid, in the form of transport grant finance; and
(iv) appendices, consisting mainly of time series statistics of transport trends.

To take one example, the TPP submission of Nottingham County Council, 1979–80 to 1983–4 shows that the general transport aim is

Table 7.14: Transport Supplementary Grant Bid by
Nottinghamshire County Council

	Current and committed expenditure (£000s)		New capital expenditure (£000s)	
	1979–80	*5 years*	*1979–80*	*5 years*
a) Capital:				
Public transport	43	52	90	190
Highways	2,483	10,817	1,717	12,456
Car parking	–	–	–	–
Lorry parking	–	–	–	–
Other	200	1,000	38	190
Total capital:	2,726	11,869	1,845	12,836
b) Current:				
Current expenditure	8,004	40,904		
Public transport subsidy	660	2,948		
Total current:	8,664	43,852		
c) Capital and current:	13,235	68,557		

Source: Nottinghamshire County Council (1978)

to strike a balance between the needs of the economy, the
environment, the pedestrian, the public transport passenger,
the commercial vehicle operator and the private car user, to
obtain the maximum freedom of movement for people and
goods in such a way that achieves a prosperous economy and a
civilized safe way of life (Nottinghamshire County Council,
1978).

From this aim, a number of more specific policy objectives were
derived, covering transport and the economy, road safety, the
environment, pedestrians, public transport, freight transport,
private transport and highways. Of these the two priorities were
'To increase the economic prosperity of the county through the
provision and maintenance of better roads and other necessary
transport facilities and infrastructure' and 'To achieve the
maximum reduction in accidents for a given expenditure through
road improvements, traffic management measures and education.'
In each priority area the TPP presents a systematic argument,
setting down: the current and future problems of, for example,
road safety; the programme (not policy) objectives, for example,
to reduce road accidents by 1.5 per cent per annum; the options

available; the proposed projects, for example, for enforcement; and a statement of monitoring, which identifies what has been happening in that particular policy area.

What this translated into, in the form of a bid for transport supplementary grant, is shown in Table 7.14.

Twenty-three further pages of forms provide the detailed costings for these proposals, and central government's Department of Transport has the responsibility to determine the actual level of grant. This it does for each of the 46 local authorities whose responsibility it is to submit a TPP. As Table 7.14 shows, the bulk of Nottinghamshire's bid was for current expenditure projects concerned with maintenance and capital projects for highway improvements. Of the total bid of £68.5 million, the Department of Transport agreed to provide grant for almost £40 million, or just over 58 per cent.

In the US, 'transport systems management plans' (TSMP) are in many respects quite similar to TPPs. Like the UK Development Plans (Structure and Local Plans), in the US all large local authorities were required to produce long-term development plans, showing proposals for land use and transport over a period of up to 20 years. In 1975 the Federal Highway Administration and UMTA published joint guidelines for much shorter-term planning, called 'transport systems management'. TSMP has been described by UMTA as a concept which involves the planning, programming and implementation of low-capital, short-range improvements designed to enhance the efficiency of existing transport systems in areas with more than 50,000 people. As Gakenheimer and Meyer (1979) have described it, TSMP tries to be more service- than facilities-orientated. The policy objectives focus on four types of project:

   (i) to make more efficient use of existing road space, such as giving preferential treatment to buses:
  (ii) to reduce vehicle use in congested areas, by supporting schemes such as restricted zones and car pooling;
 (iii) to improve public transport levels of service, by supporting express bus services; and
  (iv) to increase the operating efficiency of public transport, by stressing the need for improved marketing and maintenance.

Like TPPs, TSMP is aimed very much at systems efficiency, but also at health, environmental, equity and urban revitalisation objectives. As Paquette, Ashford and Wright (1982) show, the TSM programmes for Twin Cities and Portland resulted in the promotion of a wide range of projects, ranging from the adoption of a computerised traffic control system, the creation of bus-only lanes, the supporting of car pooling schemes (Twin Cities), the provision of park and ride facilities, and central area parking restrictions and contraflow bus lanes, to promoting fares policies, and to assisting in improving of bus stops (Portland). In each case specific policy objectives were set and objective criteria applied, and unlike the UK TPP process, most TSMPs include as a policy objective, 'to provide mobility to the disadvantaged'.

Although policy mechanisms like TPPs and TSMPs mark a significant advance in the ways in which transport policies are translated into action programmes, they nevertheless incorporate a number of quite serious problem areas which can prejudice their effectiveness. These problems are not new and in all probability are likely to surface when any new policy mechanism is introduced. The first concerns the tension created between those responsible for preparing spending programmes locally and those who ultimately control the proportion of public expenditure to go into grant aid. This is a variation on the theme of central–local relations, and is concerned with the dilemma of trying to cope with localised problems, often quite spatially specific in character, which require recognition and funding from the centre.

The second set of tensions are those to be found in all TPPs and TSMPs — the familiar problem of efficiency versus equity considerations. For example, building new roads to improve the efficiency of the highway networks in urban areas implies distributional consequences for those living in close proximity to them.

Related to this is a third tension: the conflict between using capital infrastructure projects as against providing direct income support to users, as a means of meeting mobility and accessibility objectives. For example, no amount of investment in new infrastructure will necessarily guarantee that those experiencing mobility deprivation will be able to improve their circumstances.

Finally, there is the tension between the demands associated with short-term expenditure programming and the need for longer-term development planning. In earlier decades great reliance was placed on the latter as a way of providing a strategic framework within which the former would fit. With the demise of longer-term development planning, accompanied by the rise of short-term exigency planning, there is little in the way of a strategic framework translating global transport policies into land-use/transport options. Whether this is a desirable state of affairs is debatable; what remains a constant is the need for some kind of longer-term strategic outlook within which individual investment decisions — which often take up to twenty years to complete — can at least be related to one another (Zakaria, 1985).

The Outlook

There is now clear evidence that, as structural changes in population have combined with changing economic prospects, the pattern of personal mobility has been the subject of quite major changes, both with respect to different social groups and where people live. These changes, together with a change in attitude of those responsible for informing transport policy away from mobility and toward accessibility as an organising principle, should in theory have led to a more informed and well-intentioned public policy. It is therefore perhaps quite sad that in the view of *Transport UK 1985*, a yearbook which attempts an economic, social and policy audit, the Department of Transport's scope for proposing new policies is limited, and 'it cannot guarantee that the policies it does bring in will actually work' (Harrison and Gretton, 1985). In one particularly damning section it contends that:

> The trends that argue so strongly for car and freight vehicles cast shadows over the prospects for rail and bus transport . . . The Department's forecasts have nothing to say about them or the groups who may wish to use them. Indeed, they do not even tell us the numbers of households, still less people, who will not be able to rely on the use of a car. Nor do they say anything about the nature of the trips that such households might be trying to make. Yet, there are indications that the difficulties facing those without cars will be greater than they are now.

The position of transport policy, at least in the UK, seems locked into the same partial mould which characterised its outlook twenty or more years ago: the obsession with supply-side provision rather than with users, and the dominance of road building and maintenance as its central infrastructure-based policy approach. Such an approach has little of direct relevance to offer to that 40 per cent of the population who neither use nor have access to cars, or who are dependent in whole or in part on others to provide their transport services.

With the increasing decentralisation away from large urban areas, together with an increasingly older population structure, and a concentration of service facilities in a small number of often peripheral locations, the consequences for a transport policy locked into past conventions could well prove deficient, particularly for those with or facing 'mobility disadvantage'. There are some signs of change elsewhere. In the United States, the passing of Section 16 (a) of the Urban Mass Transportation Act has placed some obligations on both those responsible for making transport policy and for providing transport services to adopt a more people-centred approach: 'elderly and handicapped persons have the same right as other persons to utilize mass transportation services' — the 'statutory rights of access' approach. A gradual change of approach in the UK is also possible in the future although this is unlikely without the generation of additional tensions.

References

Adams, J. (1981), *Transport Planning* (Routledge and Kegan Paul, London).

Bailey, J. (1979), 'Voluntary and Social Services Transport in Birmingham, Redditch and Bromsgrove', TRRL Report 467 (Transport and Road Research Laboratory, Crowthorne).

Benwell, M. (1985), 'Access and Mobility: Providing for "Special Needs" Groups', in Harrison and Gretton (eds) (1985).

Gakenheimer, R. and M. Meyer (1979), 'Urban Transportation Planning in Transition: The Sources and Prospects of TSM', *APA Journal*, January.

Goodwin, P., J. Bailey, R. Brisbourne, M. Clarke, J. Donnison, T. Render and G. Whiteley (1983), *Subsidised Public Transport and the Demand for Travel* (Gower Press, Aldershot).

HM Treasury (1982), *The Government's Expenditure Plans 1981-2 to 1984-5* (HMSO, London).

Harrison, A. and J. Gretton (eds) (1985), *Transport UK 1985* (Policy Journals, London).

Hillman, M., I. Henderson and A. Whalley (1976), *Transport Realities and Planning Policy* (Political and Economic Planning, London).

Independent Commission on Transport (1974), *Changing Directions* (Coronet Books, London).

Maggied, H. (1982), *Transportation for the Poor* (Kluwer-Nijhoff Publishing, Boston).

Moseley, M. (1979), *Accessibility: The Rural Challenge* (Methuen, London).

National Advisory Unit for Community Transport (1985), 'Study of Booking Systems' (NAUCT, Manchester).

Norman, A. (1977), *Transport and the Elderly* (National Council for the Care of Old People, London).

Nottinghamshire County Council (1978), *Transport Policies and Programme Submission for 1979–80* (West Bridgford, Notts).

OECD (1973), *Environmental Implications of Options in Urban Mobility*, Environment Directorate (OECD, Paris).

Owen, W, (1972), *The Accessible City* (Brookings Institution, Washington).

Paquette, R., N. Ashford and P. Wright (1982), *Transportation Engineering* (John Wiley, New York).

Plowden, S. (1985), *Transport Reform*, PSI 642 (Policy Studies Institute, London).

Sutton, J. (1986), *Meeting Transport Needs* (Gower Press, Aldershot).

Wachs, M. and T. Kumagai (1972), 'Physical Accessibility as a Social Indicator' (School of Architecture and Urban Planning, University of California).

Zakaria, T. (1985), 'Year 2000 Transportation Plan for the Delaware Valley Region', *Transportation Quarterly*, vol. 34, pp. 269–82.

Zeitlin, L. (1984), 'Status of Transportation Coordination in North Carolina' North Carolina Paratransit Conference.

8 TECHNICAL CHANGE AND TRANSPORT POLICY

Introduction

This book has taken as one of its explicit assumptions the notion that transport policy will, within the next decade or so, be developed within the context of existing technology. Clearly this does not mean that we have assumed away the prospects of current, state-of-the-art technology gradually evolving as advances already made, but not yet fully assimilated into the transport sector, are slowly adopted. It does mean we have assumed that no major new modes of transport will appear and that no revolutionary new forms of management will be developed. The assumption is, from an historical perspective, a reasonable one. New ideas are normally slow to be appreciated and, with the high costs of tooling up etc., take a considerable time to make appreciable inroads into the market.

As with many things there seems to be a product-life cycle with new modes of transport going through an initial period of low-volume, high-cost adoption as development and assessment takes place and before there is general acceptance and widespread use. This, in turn, and possibly after several modifications and changes on the way, leads to a final demise as technology produces more advanced alternatives or managerial drive fails. While easier to relate to tangible, mechanised modes of transport there are also life-cycles of a similar kind in management approaches and technique which are often, in practice, of equal importance. Technical change may intervene here by providing new hardware to facilitate better control or enhance the quality of information which is available to management.

One thing which is clear is that technical change in transport, and the various phases of product life cycles, are accelerating through time. While horse-drawn wheeled land transport and sail power at sea dominated transport for several thousand years, the last two centuries have seen the advent initially of, railways and, in the current century, the motor car, the lorry and air transport. Figure 8.1 provides a rough indication of the nature of this pro-

207

Figure 8.1: The Historical Progression of Transport Technology

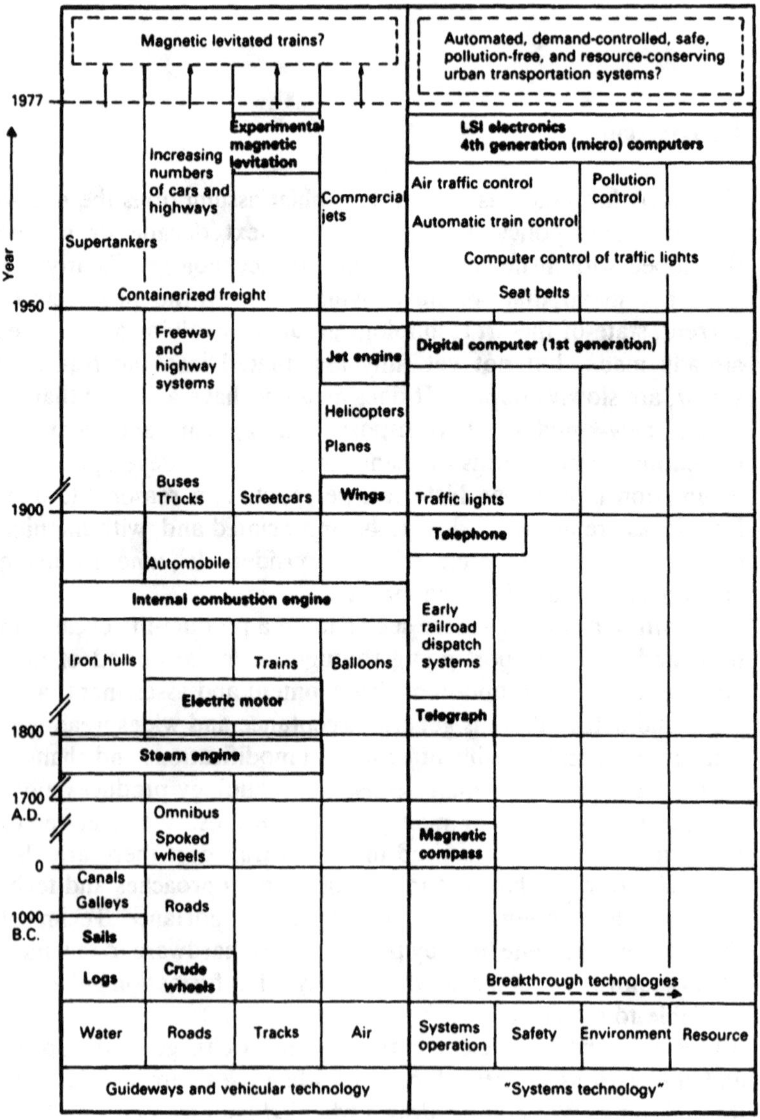

Source: R. H. Cannon (1973), 'Transportation, Automation and Societal Structure', *Proceedings of IEEE*, vol. 61, pp. 518–25.

gression together with one person's ideas of where we are likely to go over the next twenty years or so. An interesting aspect of Cannon's breakdown is the switch in emphasis which seems to be taking place in the path of technical change — a movement away from the conventional technical side towards that of 'systems technology'. In a sense this may be seen to reflect a much broader concern of many developed societies to make much better use of the transport we have rather than devote the vast majority of effort to extending the number of modes available or the scale of infrastructure provided. Obviously this does not mean that advances on what one might term the hardware side will not take place, but the emphasis is now and will continue to be an improvement on the systems side.

The shift in emphasis towards improving operational controls reflects a further point, namely that technical change is not entirely autonomous but is strongly influenced by economic and social factors. Research is conducted, by the private sector at least, in those areas where long-term financial rewards are likely to be greatest. It is sometimes possible, therefore, to glean some indication of possible technological developments by speculating on probable technical responses to major changes in economic or social conditions. We make use of this fact to a limited extent later.

Equally, it should be noted that governments (or their agents) often encourage research in the transport field, or in specific spheres within it, for reasons not directly related to transport requirements. The most obvious examples of this relate to technologies with high defence spin-offs but there are also numerous instances where prestige seems to play an important role (e.g. in the UK we may cite Concorde and the Advanced Passenger Train). In other cases research may be encouraged as a part of international trade policy (i.e. the export of technical knowledge) or for wider economic reasons not strictly related to transport efficiency (e.g. to keep people employed). Prediction of trends in these areas is difficult.

To make reasonable and specific 'guestimates' of longer term technical changes, both in the direct sense of significantly affecting the nature of the transport modes available and in the less physical area of managerial control, it is useful initially to consider some of the technical innovations which are still in their infancy but which *may* have longer term significance. For example, there have been major advances in the whole area of information technology in recent years and even the application of current, state-of-the-art knowledge on a much wider scale would produce a significant impact. This would not simply be in terms of both the direct improvement in transport performance (for example by moving closer to optimal designs of lorries, trains etc.) and management (for example by permitting a closer match of supply to demand for services), but also in terms of the nature of transport demand itself. Equally, work on electronic vehicle guidance systems and upon devices permitting vehicle identification has already yielded a technology which if applied would give traffic authorities greater control over the use made of private transport.

In other areas there is much research which, while still in its infancy (and, no doubt, there are still innovations of the past which, like the hovercraft, will suddenly emerge from obscurity when either demand conditions necessitate it or constraints on the supply side are removed), nevertheless offers some clues as to its eventual outcome and some general indication of its eventual application. New energy sources are a clear example of this with extensive experimentation going on into the use of wind and solar power as possible alternatives to the current dependence on fossil fuels. The large sums of money still going into researching potential improvements in battery design suggests that viable electric cars may also eventually materialise in the future. While not easy to extrapolate from the progress to date of such research programmes they do provide further guidelines to assist in our guestimating.

For expositional reasons it is helpful to divide our comments into a number of sections, each setting out areas where new policies may need to be formulated with respect to possible technical advances. Three broad headings prove useful in this respect, namely, Control, Energy Forms and Substitutes for Transport. We are not ambitious in what we hope to do under each heading. What we seek to do in relation to each of these topics is not necessarily to spell out in any detail the way we feel policy will develop in response to technical advances as much as to highlight the nature of the policy decisions which technological change may force onto society. The exact outcomes are much more difficult to speculate about but by trying to pinpoint particular problem areas and by examining the nature of the choices which will become available, some general impression of policy-makers' responses to technological change may be gleaned.

Control

Changes in the control of transport may take a number of different forms. At one extreme they may simply involve greater control over the information available to transport suppliers or users — essentially these are indirect instruments of control. At the other extreme, control may involve taking much of the responsibility of transport decision-making out of the hands of individuals and, instead, placing it in the cogs and micro-chips of automated systems. Technical advances in recent years have seen some movements in both these areas and, also, in control mechanisms somewhere between these extremes. Future acceptance of these changes on a large scale is likely to depend as much on social preferences and economic considerations as on the continued ingenuity of engineers and scientists.

There would seem to be fewest political difficulties in the continued development and adoption of information technology systems for transport. (The question of cost-effectiveness of systems is, of course, a different issue.) Public transport undertakings are already adopting and experimenting with a variety of systems embracing a whole range of different technologies. Radio control is now quite common while computerised bus monitoring systems were adopted a decade or so ago, for instance, in Chicago (in 1969/70), Besançon, France (in 1975), Hamburg (in 1975), Dublin (in 1975) and Paris (in 1975). More recently, London Transport has been experimenting with micro-computers (the £1.8 million 'Bisco' experiment) while New York has, since 1982, employed the CENTRAK package not just to schedule buses and to exercise operational controls but also to select optimal opportunities for bond flotations in financial markets. Some schemes are quite limited but Columbus, Ohio, with its fleet of 325 vehicles, controls them via a computerised system while the TMGTB system employed in Tokyo helps in vehicle location, office management and operational control over a fleet of 1,950 buses.

Control over private transport via information exchange is still restricted mainly to route guidance with some limited route controls (e.g. computerised systems of lights for switching lane priorities when traffic conditions necessitate greater capacity in one direction, area-wide traffic light systems etc., see Cobbe (1971) and Anderson (1977)). On-board information systems (and CB radios offer a crude form of such a system) are gradually emerging giving drivers information on traffic conditions and optimal routings. Developments of on-vehicle systems control (relating to car performance, safety and driver comfort) are also being made to help drivers make better use of their vehicles. These developments via fuel-savings systems and headway controls are also likely to yield wider social benefits relating both to the quality of the environment and the safety of other road users.

With respect to most of these topics, public policy is unlikely to meet serious social objections to the voluntary extension of such systems but decisions on common technologies and plans for systematic expansions of them are likely to be called for at an early date. Legislation to make certain types of what are sometimes called 'driver aids' (e.g. fuel monitoring systems, inadequate headway warning systems etc.) may prove more problematic but economic forces (e.g. rising fuel prices) are likely to speed up the automatic process of computerisation. In other instances, the public may need convincing of the social benefits of some of the developments (e.g. headway warning systems) as has, for example, been the case in many countries with respect to seat-belt wearing and headlight use.

More difficult are going to be questions concerning the employment of electronic identification plates (of the type currently being experimented with in Hong Kong — see Chapter 3). These offer the potential both for adopting more efficient means of traffic management and of enforcing existing driving laws (e.g. speed limits, parking controls, commercial vehicle routings etc.). Such devices, whatever their potential virtues in improving traffic conditions, are likely to be strongly resisted in some countries as they are often seen as infringements of personal privacy and a movement towards 'police state' surveillance (witness the conflicts over the introduction of mechanical tachographs into UK road haulage vehicles). Here policy-makers will need to confront rather difficult questions concerning economic efficiency and social values — the eventual policy outcome is by no means certain.

While the controversies surrounding the introduction of novel forms of information control systems are going to pose problems for policy-makers, they may not prove the most intractable problems in relation to control. A major innovation in recent years with respect to tracked modes of transport (e.g. inter-urban railways but more especially urban transit systems) and, to a lesser extent, air transport is the development of automatic control systems. These offer technologies which either remove the need for manual direction and regulation or substantially reduce the input required (see Fenton, 1970 for an early survey). Some examples of this are readily observed (such as automatic level-crossing systems, traffic lights, unmanned lifts, etc.) and have been assimilated with relatively little difficulty into the everyday operations of the transport system. Others, while at a relatively advanced stage of development, are being assimilated less rapidly — mainly because of cost considerations and public resistance rather than serious technical problems — but are, nevertheless, slowly becoming part of the accepted transport system.

If we consider these latter innovations in tracked transportation in some more detail we find that they conveniently break down into three broad types — a common feature, however, being their employment of automated, driverless vehicles. In order of current acceptance we can order them thus:

(1) At the simplest level are the shuttle loop systems (SLS) which have gradually been introduced since the 1970s. These are characterised by the relatively small, automated vehicles used and the limitations of the short loop infrastructure provided. Single loops are normally employed to minimise vehicular interaction and reduce accident risk (indeed, in the USA they have been so successful in this latter respect that only one serious accident occurred in the 1970s despite the fact that over 200 million passengers were carried). As can be seen from Table 8.1 the main applications of SLSs to date have been as short-distance person-movers in a rather limited range of circumstances. They are, however, being developed and their efficiency improved as the new system at Atlanta Airport demonstrates.

(2) While shuttle loop systems, in part because of the specific roles they are normally tailored to serve and in part because of the type of user for whom they are designed, have met with minimal objections, larger systems have been much less easy to initiate. Considerable effort has gone into the development of ground rapid transit (GRT) systems (see Table 8.2 for some international examples and Department of Transport

Table 8.1: Summary of the Characteristics of Some of the Major Shuttle Loop, Automated Transport Systems in Use in the 1970s

Installation site	SLT installation site	Year of introduction	Main tasks
Airports	Houston, Texas, USA	1969	Transporting airline passengers between landside and air terminals (at Houston airport, moreover, hotel and parking lots)
	Tampa, Florida, USA	1971	
	Seattle-Tacoma, Washington USA	1973	
Recreation parks, expositions	Hershey, Pennsylvania, USA (Amusement Park)	1969	Transporting visitors through recreation and exposition areas
	Valencia, California, USA (Magic Mountain)	1971	
	Sacramento, California, USA (California Exposition)	1975	
	Yatsu, Chiba Prefecture, Japan, System VONA (Amusement Park)	1973	
Shopping centres	Pearl Ridge, Honolulu, Hawaii, USA	1975	Linking two shopping centres
	Fairlane Town Center, USA	1976	Linking a hotel and a shopping centre
Hospitals	Ziegenhain at Kassell, FRG (System Cabinlift)	1976	Transporting persons and goods between two remote buildings
Urban cities	Lille, France (System VAL)	1975	Linking a remote university campus with the city centre and railway stations

Source: H. Sobel (ed.) (1982), *Computer Controlled Urban Transportation — A Survey of Concepts, Methods and International Experiences* (Wiley, New York).

Table 8.2: Some International Examples of Developments in Ground Rapid Transit and Personal Rapid Transit

System	Country	Development stage	Seats	Vehicle capacity Total	Speed (kmh)	Minimum headway (s)
Airtrans	USA	Passenger service since January 1974	16	40	50	18
Morgan-town		Passenger service since October 1975	8	21	48	15
Dashaveyor		Test track (demonstrated during TRANSPO '72)	12	32	48	15
Monocab			6	6	48	8
TTI			6 or 10	6 or 10	48	12
H-Bagn	West Germany	Test track	8	16	35	8
Kompakt-Bagn		Design prototype car	24	48	70	60
		Test track	12	18–20	48(72)	10(15)
KRT	Japan	EXPO '75; passenger service since 1980 in Kobe	8	23	48	15
NTC		Test track	20	50	60	20
Minitram	UK	Design	6	12	54	40–10
Characteristics of Selected PRT Systems						
CVS	Japan	Full-scale test Tokyo and Expo '75 Okinawa	4	–	60	1
CAT (Cabintaxi)	West Germany	Full-scale test Hagen in Westphalia	3	–	36	0.5 to 1
ARAMIS	France	Full-scale test Paris Orly	4	–	50	0.2 (with platooning)
Aerospace	USA	Design	6	–	32	0.5
CABTRACK	UK	Conceptual	4	–	36	0.6

Source: H. Sobel (ed.) (1982), *Computer Controlled Urban Transportation — A Survey of Concepts, Methods and International Experiences* (Wiley, New York).

(1974) for an account of US experiences) but the introduction of these would normally mean replacing existing manually operated systems of transport with consequential labour problems (e.g. the Victoria Line in London and the Paris Metro retain drivers although the systems are automated), and would require a radical change in outlook on the part of users (who, it appears, seem fearful of systems not fallible to human error).

There are also some outstanding detailed technical difficulties to overcome, relating, for instance, to software for controlling a large vehicle fleet, which at present prevent full implementation even if social acceptance was forthcoming.

(3) Personal rapid transit (PRT) is a further area where considerable resources have been committed to research. The objective here is to develop systems of high-quality transport to replace the private car for urban travel. The evidence from the work to date in this field (see again Table 8.2) is less convincing that that relating to automated mass transit modes, especially regarding the development of an acceptably cost-effective system, and, in many ways, the policy-questions we discuss below are not really likely to arise in the immediate future with respect to automated systems of this kind.

From the policy-makers' perspective, these types of development in automation, therefore, pose a series of problems. There are clear financial differences in the type of funding required for an automated system rather than manual control — generally involving higher initial capital costs but lower operating costs — and decisions as to the exact technology to favour have to be made in times of rapid technical advance. Those at the forefront of system adoption may find themselves committed to a technology which rapidly becomes obsolete. Evidence to date suggests that the interrelated question of finance and system selection are further compounded by uncertainties about performance and costs. Equally, public opposition to some changes have been encountered, often centring on concerns about safety, and this has necessitated policies designed to reassure and educate.

Possibly the most serious difficulty in the transition from manual to automatic control has been generated by the natural fears of the labour which is displaced. The replacement of men by machines takes a variety of forms and is not exclusively related to the driving control of individual vehicles. Automatic fare collection (of the kind found in Paris and in Washington, Miami and Chicago transit systems in the USA) represents a major displacement of personnel as does the adoption of electronic signalling for railways. In the future the actual direct control of vehicles is likely to become increasingly automated and labour is thus likely to become even more concerned.

In many instances compromises have been reached with labour organisations which, while having the effect of slowing the comprehensive introduction of the new technology, have also been responsible for reducing the social impact on the work force. These compromises have also helped to ease public acceptance of transport which is perceived still, albeit sometimes cosmetically, to be under the reassuring hands of manual controllers. Common ground for such compromises is, though, likely to be more difficult to uncover in the future and policy-makers may find it less easy to introduce technical innovations than in the past.

While some of the advances in automated public transit systems are already with us those relating to the public control of private modes are still very much in the experimental stage. They are, however, likely to prove a point of considerable controversy once viable and cost-effective technologies are made fully operational. Here we are talking about the gradual developments of guiding systems which would effectively remove from the driver his control over his vehicle (albeit probably only on major motorways or trunk-road links if the schemes investigated to date are carried through.) This could theoretically be achieved by under-surface electronic control systems. While central control, by optimising speed, headways etc. across the entire traffic flow, could well increase the effective capacity of the road network and improve safety they would certainly require both extensive investment programmes and major changes in public attitudes. To be realistic, it is unlikely that such schemes are going to be introduced within the foreseeable future.

If we can now attempt a brief summary of the probable impact of technological change on traffic control it would seem that there is certainly going to be a number of policy decisions to be made by the end of the century. Many of these, and certainly some of the more difficult ones, both in terms of technical alternatives and public acceptance, are likely to relate to public modes. Here the embryo of a new technology is emerging (and is ever advancing), capable of enhancing the efficiency of conventional modes, reducing the costs provision and allowing novel modes to operate effectively. The extent and timing of change is going to be important and decisions over the degree of centralisation of control will need to be made. Control over private transport is likely to continue to focus over route control but policies over electronic vehicle identification, even if they are negative, are soon going to come under review. In most instances, it seems the likely problems facing the policy-makers are going to be more of a social and economic nature than technical.

Energy Forms

The initial shocks of the two 'energy crises' of the 1970s and the follow-up ripple effects generated immediate actions to restrict fuel use, but also set in train a series of massive research efforts aimed at finding substitute fuels to replace oil. As part of this trend, there have been substantial efforts to develop more fuel-efficient transport systems. By the mid-1980s it has become clear that while some of the more dire scenarios regarding the immediate depletion of oil reserves have proved extreme there are still longer term problems associated with a continued reliance on fuel oil as the main energy source for transport. Not only are there finite stocks of oil available but a large proportion of these are located in politically sensitive areas. Also, as time proceeds and marginal deposits have to be exploited, increased geological problems will push up production costs. One must also remember that the petro-chemical industry is expanding, requiring fossil fuels for its production. Thus, while the oil companies project reserves which will meet current forecast demand to the end of the century, even if this is accurate it will only be achieved at a very high cost.

   A whole gambit of ideas exists as to what will fill both the short-term gap and the long-term needs for useful energy (see e.g. Lucas and Richards, 1982). In the short term, for instance, the available transport technology may be sustained by using oil extracted from solid carbons (e.g. coal, oil sands, etc.) but this is only likely to contain rising costs at a plateau for a short time. It is also probably going to prove a rather inefficient method of stored energy utilisation. In the longer term advances in science may provide safe means of exploiting nuclear power or alternatively 'natural' sources of energy (e.g. from the sun, tides, winds, etc.) may be more efficiently tapped. If this proves possible, then by converting the energy into hydrogen, the impact on existing forms of transport may not be very great. (Broadly, the current infrastructure could be used and the existing forms of operation continue with only minor technical modifications to the propulsion units of vehicles.)

Most of these developments or others that may materialise are unlikely to be within the strict ambit of transport policy-making. Transport is certainly a major consumer of energy (e.g. 38 per cent of petroleum consumption in the USA is by road users — 27 per cent by cars and 11 per cent by buses and trucks) but it is only one of several key sectors. Decisions within transport are, therefore, likely to be as much of a reactive nature as being strictly pro-active. Changes in the transport system will be, to a considerable extent, to conform with the nature of the sources of energy which are available and with ensuring that the optimal use is made of them. While efforts will, no doubt, be made by researchers to exploit energy sources and conventional techniques suitable to the needs of the transport industry, this will only be one dimension of their activities and transport itself will have to move some way towards transforming its operations to meet the characteristics of the energy sources available. One further point: coal, wood, petrol and other fuels account for nearly a third of the tonnage carried by transport in the world — new energy sources are themselves going to cut this and thus generate even more pressure for adaptability in the transport system.

In the recent past we could perhaps have cited an exception to this line of argument but this star seems to be fading. The hopes of the 1970s were that battery-power cars would offer a longer term solution to the problems generated by oil reserve exhaustion. Considerable research efforts were directed towards developing batteries specifically matched to the needs of transport. Such hopes now seem to be somewhat jaded. Certainly advances have been made but, firstly, the supposed environmental advantages of mobile electronic power stores have come under closer inspection and concern has risen about the environmental hazards associated with the primary generating plant (e.g. carbon-burning power stations or nuclear stations). Secondly, the advances have been disappointingly slow. The most efficient system, when examined in terms of primary energy inputs, is about 40 per cent less efficient than petrol-driven traction. (One of the main problems is the carriage of the energy itself. It takes a 50 kilo battery to contain the same energy as a litre of petrol!)

Transport policy will need, given these factors, to ensure that progress continues in research aimed at devising transport systems which are more economical in their use of current energy forms while at the same time ensuring that novel developments which have taken place are effectively assimilated into the system. One way to ensure that adequate and relevant research is conducted is to remove some of the distortive effects within the transport system so that users are fully cognisant of the energy-cost implications of their activities. This, in turn, should theoretically stimulate research into more economic technologies. In the past, for a variety of social and political reasons, this has not been the policy in most Western nations but rather attempts have been made to achieve the desired research effect by more direct means (e.g. government funding). There are also arguments that, because of the oligopolistic nature of the fuel-supply industries, coupled with discontinuities in relevant production functions, the free market could never adequately reflect the optimal guidelines for research needs.

One specific reason why many economists in the 1970s were sceptical about the power of the market to stimulate change with regard to transport fuel technology was the apparent insensitivity of users to the price of fuel (an idea perpetuated by early econometric studies which revealed very low fuel price elasticities of demand — found to be as low as 0.06). If fuel prices rose as reserves were exhausted, the argument went, this would be absorbed by the willingness of transport users to pay rather than stimulating research initiatives for more fuel-efficient technologies on the part of the transport suppliers. More recent work using both improved estimation procedures and better data sets covering longer adjustment periods has shown that, rather than being insensitive to fuel price, consumers' demand is in the longer term highly elastic (in the order of 1.1 to 1.9). Once they are sure the price rise is permanent, then transport users seek out smaller, lighter and less fuel-hungry vehicles. Making users fully cognisant of the economic price of petrol, therefore, provides both a stimulus for private-sector research into new energy forms and enhances public support for state research initiatives.

The emerging empirical evidence is that at least in some areas financial incentives generated by the higher oil prices since the early 1970s are bringing forth improved technologies. The jolt that the sudden escalation in oil prices inflated on the market seems to have been sufficient to have overcome any imperfections inherent in the market. Even at the anecdotal level the casual observer cannot fail to have noticed the improved fuel economy advertised by car manufacturers and the emphasis now put on 'slippery' bodies and alloy components.

The power of the market, therefore, clearly seems to stimulate technical change over time wherever shortages emerge but equally, as we suggest above, the mechanism, especially in the short term, may prove somewhat sticky. The ability on the part of key actors to exercise monopoly power is the most obvious cause of friction but there is also the fact that the producers and consumers have to be convinced that any price rise is permanent. Given this situation policies have often involved and are likely to continue to involve more immediate government directions. An illustration of this is once more provided by the 1970s oil crises which stimulated the USA to initiate legislation in 1978 requiring car manufacturers to improve progressively the fuel economy of their products. The econometric evidence to date (e.g. Munro and Schachter, 1982) suggests this programme has proved successful in encouraging fuel economy. Of course, there is no guarantee that the progress is actually optimal (greater emphasis may, for example, have been more beneficially devoted to longer term programmes of research rather than to the meeting of short-term targets) and care must be taken in specifying such forms of directive but in emergency situations where, for one reason or another, the market may be sluggish, they have been employed with effect.

While most attention has focused on the energy demands of private land transport, less obvious, but nevertheless important, is the concern now being shown by manufacturers for greater economy in fuel consumption to accompany the new generation of commercial aircraft. (The first licensing of commercial airships in the UK in 1984 and their gradual adoption in Canada for timber carriage are extreme examples of the direction of events here — their real impact may come in the next century.) Shipping has also not escaped and not only are new vessels being equipped with less thirsty engines but ideas of utilising supplementary wind power sources are being mooted (and, indeed, even being adopted on a small scale). These are all essentially developments brought about by economic pressures rather than government regulation. Indeed, the removal of government regulations in the USA demolished shelters enjoyed by the domestic airline companies and forced them to seek great economy in the face of high fuel bills.

In summary, it would seem probable that government will allow greater market freedom with respect to its goals of both developing new energy sources to propel transport and to stimulate their efficient adoption. If for no other reason, this will keep the maximum number of options open. In the short term, although possibly with greater care than some countries have exercised in the past, resort is likely, and legitimately so, to be made to more direct policies. Intellectually the latter may be justified in terms of the greater amount of current information normally available to government than to individual suppliers but more practically it satisfied public concern that government is seen to be active in a field that is perceived as important to both current and future generations. Within this framework of the free market dictating longer term trends but including short-term injections of government activity, there is also likely to be increased concern, especially in the medium term if the majority of forecasts are correct, about ensuring adequate mobility for the less affluent during a period when new energy sources are developed and existing ones become more expensive. Energy policy in the transport context is, therefore, likely to require a much more coherently thought-out distributional element than has been the situation in the past.

Substitutes for Transport

Technical progress in the transport sector itself, as we have seen, is likely to pose problems for decision-makers but equally there are appreciable changes taking place outside transport which are likely to have repercussions for that sector. People use transport not for its own sake (although there are, of course, a few joy riders) but rather to attain some other aim, be it in terms of production or their own immediate well-being. The types of technical change we are concerned with in this section involve the development of means of attaining these final objectives without the need (or at least in some cases with a lesser need) for transport services. Essentially we are looking at what are sometimes called 'knock-back' effects.

The substitution of something else in place of transport is not new — throughout time people have tried to avoid transport. Examples abound where technical change has actually done away with transport or substantially reduced its importance in the movement of goods. (The refrigerator removed the need to deliver ice by wagon; fruit concentrates have reduced the need to transport peel and pulp around the country; gas and electricity have reduced the transport required to move coal-extracted energy, etc.) Perhaps the major innovations are now not going to be so much in terms of goods movements and the consumption of goods but rather relate to services and personal travel.

Information technology is now seen as the main alternative to transport in some key spheres, especially those relating to service sector activities. As we have seen (in Chapter 6), technical change has been altering the requirements of modern manufacturing industry but equally there are now in train significant technical advances which may well influence both the nature of many services that people use and the transport required to obtain them.

The advent of the micro-computer together with the ability to link individual systems is increasingly offering the potential for automatic exchanges in information and subsequent reductions in the number of personal transactions. Already in the UK some financial institutions are providing customers with micro-computers capable of conducting transactions with the main-frame machine over telephone lines, thus cutting out the need for office visitations. It is technically possible now for households to shop via computerised systems and then to receive home delivery of the goods — once more reducing the number of personal trips involved and also reducing the number of movements of goods in the selling process. Home entertainment is becoming more sophisticated and interactive games etc. are gradually reducing the need to leave the home to enjoy competitive forms of participation.

In the sphere of business the pace of change is even more rapid although, to date, the impact on aggregate travel demand is less clear. While high technology has reduced the need for many trips and permitted the rationalisation of others into fewer, shorter movements it has also stimulated new forms of activity which at present are still being assimilated. Taking a view of the future it seems that once the transition is more fully under way then transport needs will decline in an absolute sense. The ability to store information and process it when necessary is also likely to permit greater flexibility in work patterns — it is likely that this will affect the timing of trips as well as their magnitude.

We should also remember that many of the changes which are and which will take place will relate to improvements in technologies we already have. Third World cities, for example, suffer from the inadequacy and unreliability of their telephone systems. This results in very high numbers of personal meetings and, *ipso facto*, much cross-city transport. New technologies are improving the reliability of telephone and related communications systems and, of equal importance, reducing their real costs. Technological advances can, therefore, have major implications even when they do not radically alter the outward nature of the types of communication currently available.

The problem which is going to confront policy-makers is to incorporate the likely longer term impacts of the current informational technology revolution into their short- and medium-term policies. Current needs are still largely influenced by past patterns of demand which, in turn, are formed by the declining sectors of the economy and by the social habits of the pre-information technology society. Excessive investments to meet these specific needs could prove wasteful if major changes in travel behaviour do take place. This would suggest that a priority on flexibility is required in the types of policy which are adopted. The difficulty is that the availability of real substitutes for transport represents a dimension to the transport policy debate which did not exist before — it is not just a change in the magnitude of demand but may reveal itself as a major break from previous trends.

Some Closing Comments

As we made clear at the outset, we have not been concerned with stargazing in this chapter but rather with trying to predict the types of transport policy-decisions which will have to be made in the future in the face of emerging new technologies. Our coverage has not been comprehensive but has concentrated on three major spheres where technical changes seem to us likely to have an impact on policy-making.

Of course, even the limited discussion that we have offered is all extremely speculative but one or two points do seem to stand out. Briefly these involve the difficulty of trying to predict with any certainty exactly how existing technology is going to develop in the immediate future, let alone what entirely new developments are going to take place. It seems unlikely that any new mode will have the impact of the railways in the mid-nineteenth century but changes of a lesser order (such as the introduction of wide-bodied jets in the 1960s and 1970s) can still produce mini-revolutions. Rigid planning on the basis of either existing technology or on the belief that some new, specified technology will emerge is, therefore, extremely dangerous. This seems to be a fact with which policy-makers are now coming to terms. Flexibility and pragmatism are replacing the tendency towards network planning and comprehensive regulation which characterised many countries' transport policies until comparatively recently.

Equally, it is clear that transport cannot be treated in near isolation from other industrial sectors as has often been the practice in the past. It has traditionally been perceived that transport will change in response to changes elsewhere in the system but that its role will essentially remain the same. Recent advances in information technology (coupled with major shifts in the nature of most Western economies) means that alternatives to transport are increasingly appearing which may be adopted quite rapidly to replace transport inputs. Policy, to be effective, is almost certainly going to have to become more sensitive to other areas of economic and social activity than it has been in the past.

References

Anderson, P. E. (1977). 'A Revolution in Electronics', *Traffic Engineering*, vol. 47, pp. 17–18.

Button, K. J. (1982), *Transport Economics* (Heinemann, London).

Cobbe, B. M. (1971), 'Computer Control of Traffic', *British Symposium of Advanced Technology* (BNEC Export Council for Europe and the London Chamber of Commerce, Zurich).

Department of Transportation (1974), 'United States Department of Transportation Automated Urban Transportation System Developments', *Report from the US Department of Transportation* (Urban Mass Transportation Administration, Washington).

Fenton, R. E. (1970), 'Automatic Vehicle Guidance and Control — A State-of-the-Art Survey', *IEEE Transactions on Vehicular Technology*, VT–19(1), pp. 153–61.

Lucas, G. G. and W. L. Richards (1982), 'Alternative Fuels for Transportation', *Transportation Planning and Technology*, vol. 7, pp. 167–70.

Munro, J. M. and G. Schachter (1982), 'Energy Policies in the Transportation Sector in OECD Countries', *International Journal of Transport Economics*, vol. 9, pp. 291–306.

INDEX

For Product Safety Concerns and Information please contact our EU
representative GPSR@taylorandfrancis.com
Taylor & Francis Verlag GmbH, Kaufingerstraße 24, 80331 München, Germany

www.ingramcontent.com/pod-product-compliance
Lightning Source LLC
Chambersburg PA
CBHW050425280326
41932CB00013BA/1997